TORHOUSEMUIR:

MEMORIES OF A WIGTOWNSHIRE CROFTER 1935-1945

Joe Whiteford

Edited by Julia Muir Watt

Illustrated by Chris Sandbach

Dumfries and Galloway
Libraries, Information and Archives
2001

First published 2001
© publication copyright Dumfries and Galloway Council
© text copyright Joe Whiteford

Design, set and print by Solway Offset Services Limited, Dumfries for the publisher, Dumfries and Galloway Libraries, Information and Archives

Torhousemuir: Memories of a Wigtownshire Crofter
1935-1945
Joe Whiteford
Edited by Julia Muir Watt
Illustrated by Chris Sandbach

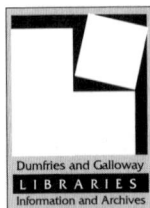

ISBN 0 946280 54 1

**Dumfries and Galloway
Libraries, Information and Archives
Central Support Unit,
Catherine Street
Dumfries DG1 1JB**

Dumfries and Galloway
LIBRARIES
Information and Archives

We also publish -
Dumfries and Galloway : *Through the Lens*

Wigtownshire	ISBN 0 946280 21 5	Machars Farm Life	ISBN 0 946280 28 2
Dalbeattie and District	ISBN 0 946280 25 8	Whithorn	ISBN 0 946280 23 1
Thornhill and District	ISBN 0 946280 25 8	South Machars	ISBN 0 946280 24 X
Queen of the South	ISBN 0 946280 27 4	Sanquhar and District	ISBN 0 946280 29 0
South Rhinns	ISBN 0 946280 32 0	Langholm and District	ISBN 0 946280 30 4
Stranraer F C	ISBN 0 946280 37 1	Characters of D and G	ISBN 0 946280 36 3
Dalbeattie Words	ISBN 0 946280 34 7	Maxwelltown	ISBN 0 946280 40 1
Creetown	ISBN 0 946280 41 X	Whithorn and Glasserton	ISBN 0 946280 42 8
Galloway Seaports	ISBN 0 946280 47 9	Ruthwell, Cummertrees	
Eaglesfield, Kirtlebridge, Middlebie		and Mouswald	ISBN 0 946280 45 2
and Waterbeck	ISBN 0 946280 48 7		

The titles above are available through local libraries or any good bookshop, priced £2.50 or £3.50 per title including postage and packing from the above address.

Other titles available include-

Ross's Story: 20th Century farming in South West Scotland	ISBN 0 946280 31 2
Sang o the Nith: a poetic journey through Dumfries and Galloway	ISBN 0 946280 19 3
From the Serchio to the Solway	ISBN 0 946280 33 9
The Sound of our Voices, Dumfries and Galloway poetry, prose and photography	ISBN 0 946280 39 X
Chinese Spare Ribs, young poets of Dumfries and Galloway	ISBN 0 946280 35 5
Dumfries and Galloway a literary guide	ISBN 0 946280 46 0
Dumfries and Galloway a literary map	ISBN 0 946280 50 9

A complete list of our publications is available from the above address or on our website at www.dumgal.gov.uk/lia. Our e-mail address is libs&i@dumgal.gov.uk

INTRODUCTION

AT first sight, the sub-title *Memories of a Wigtownshire Crofter* may seem to present a contradiction in terms. Crofting and the crofting landscape have become, almost by definition, part of the history and identity of the Highlands and Islands, whereas Wigtownshire is the most southerly county in Scotland, and, from the nineteenth century onwards, has been particularly identified with large scale dairy farming. The Highland connotation of the word *crofting* has been enshrined in legal form, from the Napier Commission of 1883 to the Crofters Commission of today, whose responsibility is to *organise, develop and regulate crofting in the crofting areas of Scotland, viz. the former counties of Shetland, Orkney, Caithness, Sutherland, Ross and Cromarty, Inverness and Argyll...*

This appearance of paradox is the reason for this book: to document, from the living memory of a Wigtownshire crofter, the existence of small-scale landholdings in the extreme south of Scotland, which are now only known through a name on a map, or from some puzzling archaeological remains in a landscape of intensive farming. Torhousemuir may have been an anomaly or even a fossil sixty-five years ago, but its existence and its possibility are facts which demand all the more recognition and recording.

Necessarily, there are differences from the northern crofts: the Torhousemuir crofts are not set in a rugged landscape of sea and mountain, but rather in a mossy hollow, surrounded by drumlins of boulder clay, in the low-lying Machars peninsula. By the time the Whiteford family arrived at Torhousemuir in 1935, Wigtownshire, with its mild pasture lands, was well-established as a supplier of liquid milk to the cities, and the monthly milk cheque gave the crofters a cash income which was denied to the Highland crofters, struggling on the edge of self-sufficiency. Nonetheless, the crofters of Torhousemuir survived on acreages well within the definition of crofting, and held their land on individual tenancies on a basis typical of crofting. While the land of the Machars is not considered marginal like the traditional crofting areas of the north, the holdings of Torhousemuir were nonetheless on peripheral land, and Dr. Richard Oram has argued, in his report[1] on the settlement, that the evidence suggests that it was formed by the landowner in the nineteeth century in order to exploit areas of poor

---

1   *Torhousemuir Historical Account. Report on Programme of Research undertaken by Retrospect Historical Services on Behalf of Scottish Natural Heritage, February 1995. Richard D. Oram, MA PhD.* I am grateful to Dr. Oram for his permission to refer to his research in this publication.

land on his estate. Notoriously, crofting in the north too had its origin in the social and economic experiment of the Clearances and subsequent re-settlement of families on marginal, often coastal, land, which they were given the task of reclaiming for agricultural use. While the Wigtownshire crofters rarely did work outside the croft (except perhaps for helping neighbours), unlike their Highland counterparts who depended partly on fishing and ancillary trades to enable them to survive, it is also evident in the following account that life at Torhousemuir was also an existence at the very edge of the cash economy. The crofters survived economically – those who did, that is – because the expenditure of cash was kept to an absolute minimum: the use of the crofters' own labour, including the labour of the entire family unit, the ethic of self-sufficiency, saving and repair, the hiring out of labour beyond the croft, and the cultivation of a mode of agriculture which produced seed and stock for the following season, all contributed to a conservative existence in which the money economy and technological advance were kept at arm's length.

THE TORHOUSEMUIR ESTATE

A brief reference to the history of the estate may help locate the particular late phase of its development which concerns us here; for details, the reader is referred to Dr. Oram's report and to M'Kerlie's *Lands and Their Owners in Galloway*. The Torhouse estate came into being upon the division by the Crown of the former Wigtownshire lands of the lordship of Galloway. While the burgh of Wigtown became a royal sheriffdom, the estate was held independently – originally by Hugh de Turfhouse sometime in the mid thirteenth century. In the fifteenth century, when the estate of Torhouse was in the possession of the MacDowell family, the divisions which survive to the present were created, taking their names from the spouses or tenants of the heiress, Affrica MacDowell: Torhouse proper (Torhouse-McCulloch), Torhouse-M'Kie (now Torhousekie) and Torhouse-Mure (from the Wigtown family of Mure).

Archibald Mure left the property which we now know as Torhousemuir (also known at Balmeg) to his three daughters in 1543. The Gordon family held it until the late eighteenth century, and it was then sold by them to the Thomsons, a Stewartry family. In 1827, Captain James McHaffie of Fuffock acquired the property and it remained in the family until 1908. By then, it had begun to exchange hands more rapidly: a Staffordshire landowner, John Stubbs acquired it in 1908 and sold it on to a nephew, Colonel Arthur Hall. In 1934, just a year before the Whiteford family moved to Mossend croft, Lord John Fitzroy acquired it, and sold eventually in the 1943 to the White family, whose descendants are still in possession of the estate.

From early nineteenth century maps, it becomes clear that the form of the estate which existed in the 1930's had come into existence approximately one hundred years before. In the 1830's, the two substantial farms of Meadowbank and Knockmore were already in existence, surrounded by a patchwork of 38 smallholdings. The enclosed oval which bounds the crofting settlement probably pre-dates the sub-division by at least sixty years, according to Dr. Oram, and was probably constructed in the era of agricultural improvement and enclosure of the late eighteenth century. By the 1930's, the numbers of crofts had been reduced by a process of amalgamation from 38 to twelve, but the ruins of formerly independent crofts were then (as now, in some cases) still visible.

For the historian, the problem is to account for the process of sub-division on Torhousemuir estate, at a time when the forces of agricultural improvement were driving owners and tenants in quite the opposite direction. Within Wigtown parish, as elsewhere, the way in which Torhousemuir was developing was clearly unusual and was singled out for comment by the compilers of the 1841 Census[2]. An analysis by Dr. Oram of the demographics of Torhousemuir suggests that the majority of early settlers were migrants from Ulster, and that the landowner, James McHaffie, may have taken advantage of the pool of land-hungry Irish to develop marginal land by offering improving leases, and may even have advertised the plots directly in Ireland. There may therefore have been, underlying this intensive subdivision, an element both of social experiment and of economic self-interest by the landowner.

By the late nineteenth century, when James McHaffie's son, William, took over in 1865, the numbers of occupied crofts had shrunk to about fifteen and there is evidence of an increased number of ruined or unoccupied buildings. This may have been owing either to the failure of the crofts with poorer land (closer to the moss in the north of the estate) or due to the success of some of those with better land, who could afford to acquire neighbouring plots. Interestingly, however, the number of crofts in active occupation by the end of the nineteenth century stayed relatively stable into the twentieth, since twelve were still being farmed by the time the Whitefords arrived at Torhousemuir. One can, however, see the process of

---

2 See Oram, *op. cit.*, page v.

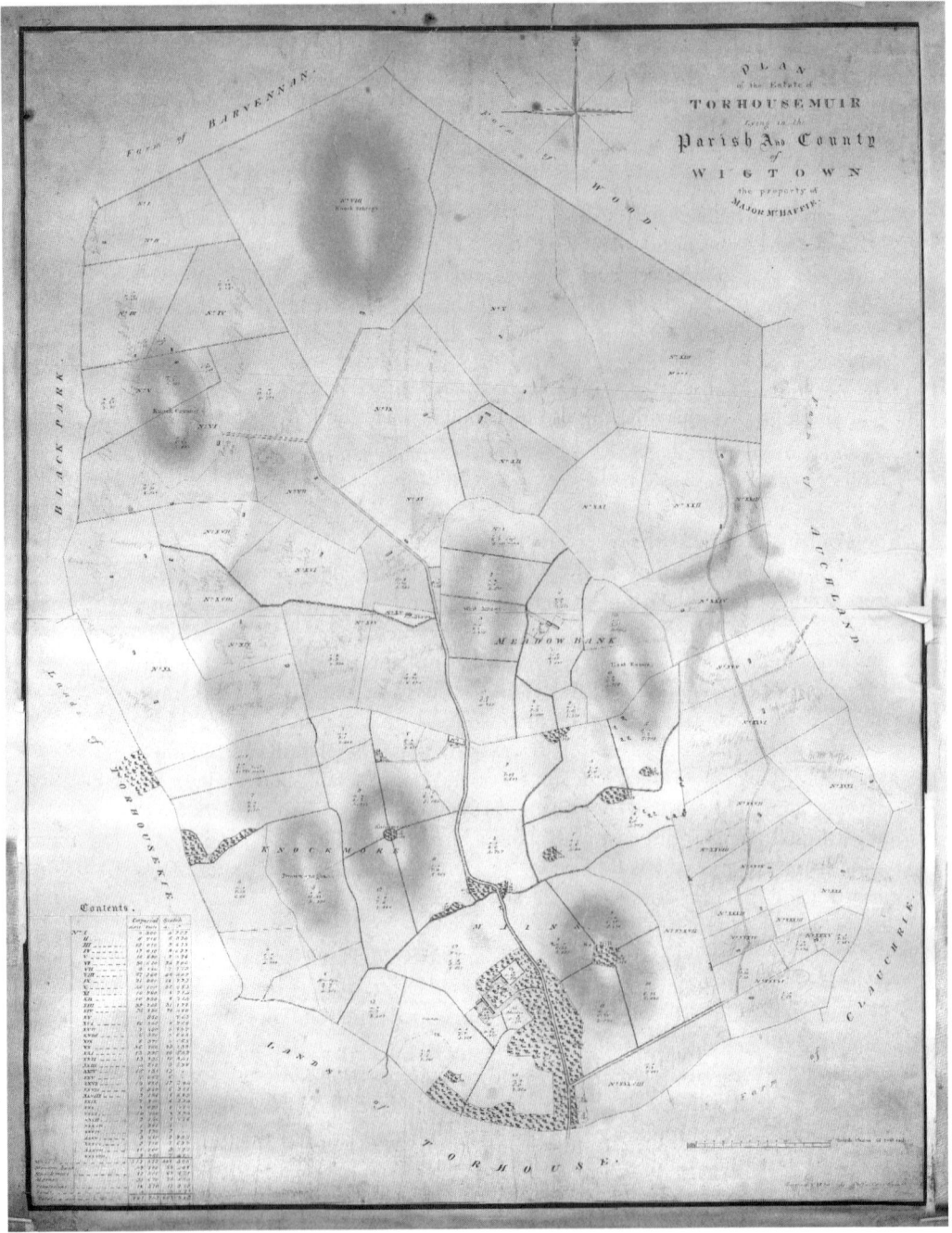

Torhousemuir from a survey by Colin Christison, 1834,
kindly lent by Mr. & Mrs. Orr Ewing of Torhousemuir.

amalgamation speeding up through the twentieth century, from the period of the witness account which follows, to the present, when all but one (Knockskeog) of the original crofts have been absorbed back into the direct holding of the estate, and when most of the buildings still occupied in the 1930's have been abandoned.

AGRICULTURAL TECHNIQUE

The value of the following account of life at Torhousemuir, then, is its documenting of a way of life which was unusual for its place and anomalous for its time. In the 1930's, not only did its physical layout reflect the planning of a previous era, but also the small acreages and low capitalisation held back investment in agricultural technology. Whereas elsewhere in Wigtownshire there was evidence of mechanisation (albeit horse-drawn), at Torhousemuir practices such as broadcast sowing and *opening the roads* with a scythe were preserved on the crofts, because of the crofters' need to maximise use of their own labour and to avoid waste of any description. In some respects, therefore, the practices on the crofts were anachronistic for the time and can give us insight through eyewitness testimony into a much earlier era; to many, even those involved in farming today, such practices will be unfamiliar. For these reasons, the detailed descriptions of farming practices have been given prominence in the following account. While there are many books on farming bygones, to find descriptions of what it was actually like, at first hand, to plough or sow, thatch a stack, or to have explanations as to why and with what purpose stacks were built in a certain way, and what role each practice had in the economy of the croft as a whole, is a relative rarity.

In another sense too, these memoirs are unusual, simply because they document a success story. The history of the crofts and an examination of the turnover of tenancies proves that success on this land under these conditions was not easily won. While in the 'thirties, most of those at Torhousemuir were of retiring age and, of those younger families who came, many gave up their leases early, the Whitefords proved the exception to the rule. Existence on the croft was poised on a knife-edge, where waste or an unwise expenditure of cash where there was an alternative, could prove ruinous. Those who observed the unwritten rules of crofting existence, however, and who were presumably driven by the desire ultimately to lease a larger holding, or perhaps even to own land, could apparently succeed, despite the poor land and small acreages. The Whitefords were one of a small group who were ultimately able to garner enough resources to stock a larger holding.

The decade to which these memoirs refer was also exceptional: on the eve of the Second World War, some the techniques being practised in the croft probably harked back two hundred years, others still to medieval times. From 1945 onwards, an entirely new world of manpower and motive power quickened the rate of change, vastly increased the capitalisation required on farms and decreased the significance of what the crofter had always been able to rely upon - himself and his skills. The bulk tanks for milk and milking parlours, hygiene regulations, the development of new crops which made old rotations redundant: all these contributed to undermining the crofter's ability to compete using his labour alone. One senses, as the sub-text of the memoirs which follow here, the gathering – by 1945 - of all the forces which were to make crofting in Wigtownshire impossible.

It is no part of this book to argue for the desirability of one mode of agriculture over another: this is simply an unvarnished account of what actually happened at a particular time, under particular, if untypical, conditions. Nonetheless, it is impossible not to note from the bare description of farming technique the way in which the crofters were far more exposed – physically and economically – to the processes of nature than is even thinkable now. Without the technological dominance offered by machines largely indifferent to the vagaries of the weather, and without strains of crop which overcome a natural disadvantage or unsuitability of soil or climate, the revolving of the seasons had a life-and-death significance which we can no longer sense. This process of internalisation – one thinks even of the construction and comfort of houses, the cabs which came to cover tractors, and in general, the layers of security which come to be interposed between man and his environment – cuts us off from the experiences narrated in the following pages. It is not simply time which separates us from this former life, but a cocooning which insulates us from its directness and makes the familiar unfamiliar.

But if the rawness and, as it were, the nakedness of pre-mechanised farming strikes one, so too does the calculatedness of its technique. To equate low technology with a lack of precision is simply to show the city-dweller's ignorance of detail: we are told that a skilled man could calculate the fertiliser or the seed required for a certain field to within ounces of the correct total. But this was not done following instructions on a packet calculated in a laboratory elsewhere: it was calculated according to a much more readily available rule - to the tread and reach of the sower, who knew his capacities and his field. The sowing and the harvest were in turn calculated in accordance with the load, a consideration of the horse and the cart, and the labour of reaping; the timing of harvest and the storage of its fruits were planned with an eye to the oncoming winter and a prudent provision for the following year. Measurements were certainly weighed up, but the bearings from which they were taken belonged to a smaller world than they do now: what was possible or impossible were hemmed in more closely by what the elements would allow, time more tightly constrained by the seasons and calculated by the capacity of a man working his horse. To all this, the bleak bass-note was the imperative for the family to survive and labour the following season.

The relentless nature of the cycle in agriculture has become a truism in writing on the countryside, but today, when there are ever greater possibilities of action at a distance, of systems which work without direct human intervention, it seems worth stressing the constant and urgent burden of daily tasks from which there was no escape. Little wonder, then, that some Torhousemuir crofters rarely, if ever, left the estate. Little wonder, either, that in this face-to-face world where word of mouth brought news, where communication was slower and less all-penetrating, sayings were repeated, curiosity about one's neighbours flourished, and habits settled into eccentricities.

An entire mentality of conservation and re-use grew up with the economic necessity of preserving cash: dung was ploughed into the fields, jumpers were re-cast as socks, flour bags transformed into trouser linings. Each thing survived into a second and surprising life, transformed for another use, eked out into another existence. Perhaps the supposedly characteristic Scots penny-pinching, which has been the butt of innumerable anecdotes and jokes, has actually been a specific product of specific conditions, with their own necessity and logic – ones which, along with the objects and fabrics suited to such re-use, in fact no longer exist. One can glimpse the fundamental anxiety for survival in the respect paid to the *skilled man* and the corresponding (almost moral) condemnation of bad or lax practices, which might tip the delicate balance of the croft towards debt or ruin.

Perhaps in this context it is worth noting that the entire account can be read as a documentary footnote to the writings of John MacNeillie, who later wrote as Ian Niall. The road on which Torhousemuir is situated also leads to the childhood home of Niall, the period of these memories is only marginally later than his own and some of the characters who appear briefly here were known to Niall. In all respects, the period, the technology and the landscape are vintage Niall territory: not only the geographical location, but the lost world it represents.

METHOD

It is important that the method used here should be understood, so that the book can be read as a reliable first-hand account . The text is based on approximately one year's discussions between myself as editor and Joe Whiteford as informant. The sessions were based on questions and answers, but would generally turn into longer descriptions of a particular point or aspect as our focus developed on that particular occasion. The whole text was then edited together into a cohesive whole, with different sections concentrating on particular aspects of life at Torhousemuir. While the editing was necessary in order to bring together disparate remarks which might have been made many months apart, each sentence in the text reflects a spoken comment and has not been introduced by the editor. The headings are elements which were inserted by the editor, for ease of reference. It was subsequently re-read by Joe Whiteford and any amendments he felt necessary made. The decision was taken early on to illustrate the book fully, given the difficulties which I myself had experienced in understanding some of the technicalities of crofting practice. Given the absence of original photographs, we are grateful to Chris Sandbach for re-creating the world of the croft in his drawings.

This is not a work by or for sociologists, but there are nonetheless insights which can be derived from it. There are many hints, not necessarily made explicit in the text, about the respective roles of men and women on the croft, the cohesiveness of the family unit, (particularly when it came to the support of the family by sons living away from home), attitudes towards religion and education, hygiene and cleanliness, privacy and sexuality, leisure and personal expectation. If these attitudes are strikingly different from ours, then this book has done its work in recording something unique, now past, and in sketching in a broader background of possibility, against which our own sensibilities and prejudices can also be seen as themselves conditioned by particular factors and technologies, and as belonging specifically to their own historical landscape.

TORHOUSEMUIR:
Memories of a Wigtownshire Crofter
1935-45

PRELIMINARIES

MY first memory of Torhousemuir is of waking abruptly when the lorry, carrying all our household goods, gave a lurch: all that was to be seen was a small gate through a stone dyke, in the glare of the headlights. It was December 1935 and I was five years old.

In fact, we weren't at the end of the road at all; there was still a mile to go, up the long road to Mossend, which I came to know so well. It was simply that the lorry had given a particularly vigorous jolt, probably negotiating the sharp bend, and this was the first time I had wakened on this journey from my first home at The Ross, in the Stewartry of Kircudbright, to our new home and first croft at Torhousemuir. All I knew was that everyone was in a bad mood: the move had not gone smoothly. The haulage contractor, Alec Love, had got drunk and failed to turn up for the removal until after three o'clock in the afternoon. This meant that, after loading, we would not be at Torhousemuir before eleven o'clock at night. My mother and I sat up in the wagon, while my three brothers and my sister were on sofas in the back under the canvas stretched over the load.

When we did arrive at Mossend, as our croft was called, it was 11.30pm. and, not surprisingly, the outgoing tenant refused to leave until the next day. We stacked the furniture in the barn, and, while my mother, sister and I slept in a spare room, my elder brothers bedded down in the barn. My father was not with us: he had to

Alec Whiteford, Joe's father.

work out six months' notice at The Ross, and did not move until May 1936, when he brought with him my brother Jimmy who was at school in Borgue. In all, there were eight children in the family, of which I was the youngest. Some were already grown up and working by the time we arrived at Torhousemuir.

I said that the Ross was my first home, but in fact my parents had moved about a great deal, and I remember thinking that my father would not get a long service medal as either ploughman or shepherd, which were the positions he usually applied for. The Ross was simply the first home I can remember; but even in my short life, I had already moved five times. I was born at Penninghame in 1929, moved to Inch at Sorbie, and then to the neighbourhood of Borgue at Balmangan. There we lived in a farm cottage called Castle Rennie. We moved again to Brighouse, and then to the Ross, where my father was employed as shepherd. I once asked my mother why we moved so often – that is, until we reached Torhousemuir –

and she said that if a move brought an extra two shillings a week, it was worthwhile. I suppose, with a family which eventually grew to eight children, the opportunism was understandable.

My earliest memories are of The Ross. The farm is situated on the tip of the peninsula, and includes Ross Island, and I can remember my father, brother and Mr. Finlay, the farmer, rowing sheep from the island back to the farm. I can also remember funerals having to pass our cottage on their way to Senwick churchyard, along a track which is now hardly visible. The coffin used to pass, balanced on a cart, and I can remember being dragged indoors when peeping at it from behind a gatepost. The other thing which sticks in my mind and must have astonished me at the time, was my first view of otters, which were basking in Senwick churchyard only yards from the sea. I began school at Borgue, and we were taken there by Raffle and Rogerson, the motor-hirers.

While we were at The Ross, two of my sisters – Jean and Min - left home for London. They had been *in service* at Galloway House, Garlieston, with the Forteviot family. Their stay in London did not last long: our uncle, a draper in Seven Sisters Road, found them in poor accommodation in Maida Vale, and after this, Jean returned to the Ross and, after our move to Torhousemuir, stayed at The Ross looking after my father and brother Jim, until the notice at the farm was worked out. Minnie became a child minder with an Indian doctor until the War, when he returned to India. I can remember her bringing his son to Torhousemuir during the holidays. Eventually, she became housekeeper to John Charles Lamont, at the Stell, Kirkcudbright, and herself got married.

John and Alistair, my elder brothers, made the move with us in November 1935, and worked on neighbouring farms, to help supplement the family income. My father was Alec Whiteford, from Bishopton near Whithorn. He was ploughing at Bishopton by the age of fourteen. He used to return to Whithorn on New Year's night, staying overnight and returning to Torhousemuir the following day. My mother, Maggie Templeton, was an Ayrshirewoman, born in Cumnock and brought up in Auchinleck; she met my father when he was working in that neighbourhood. The first house they occupied in their married life was called *The Canteen* on the farm at Garrarie, near Whithorn, probably some time around 1918, and my oldest sister was born there. Another sister was born at Brownhill Cottage, near Garlieston. Other moves within the South Machars followed: to the Craig, to Appleby, to Portyerrock, to High Balcray, and then to Penninghame, where I was born.

The renting of the croft at Torhousemuir was a different sort of move, however: first of all, we were there for ten years, and secondly, it meant that we had a place of our own. Crofters might be a step below farmers, but they were a step above cottars and farm workers. Renting the croft meant that we had a foot on the ladder towards owning our own land and I am sure my father knew it. By 1945, at the end of our ten year lease of Mossend, we did move to a bigger croft at Torhousekie; eventually my brothers came to own the house there. You tended to find, in fact, that those who waited out their leases did go on to better things, but there were many who did not make a success of it and left with the *break* at five years.

All in all, though, Torhousemuir in 1935 looked like an unlikely place to begin a rise in fortune: most of the tenants were old, or at least seemed very old to us children, and the ground was rough, though not as rough as it looks today, now that the ground is not cultivated and the crofts are abandoned. Perhaps it was just those poor conditions which made my parents more determined to succeed and more capable of it. Certainly, my father was sought after as a competent worker on the farms. Both my parents were both quiet, self-contained people: probably years of working in the fields on his own or with family members made my father a silent, contented sort of man, seeking very little in his leisure time beyond the company of his pipe. My mother had a very placid disposition, and it seemed nothing could upset her. It took someone like my mother to move into the croft and set it up, in the absence of my father, as a going concern: she set immediately about the buying of cows and calves to bring forward, travelling by herself to Whithorn to negotiate with a Mr. Hughes. Along with her work on the croft, she also managed to rear eight children. On reflection, in all those years which followed at Torhousemuir, I never did once see my mother idle.

THE ROAD TO MOSSEND

TORHOUSEMUIR road-end is about three miles from Wigtown, in the Machars of Wigtownshire. You find it on the Kirkcowan road, which runs from Wigtown crossroads, more or less parallel with the river Bladnoch. Torhousemuir was the name of the estate and the big house, but it was also given to the group of crofts which the landowner had built and settled on the land north of the house. Beyond the road-end on the Kirkcowan road, it is half a mile before you reach the estate and the big house, and a mile beyond that, you reach our croft of Mossend. The estate was surrounded by moor on three sides - Clauchrie, Auchleand, and Blackpark Moor – and it was not generally good land.

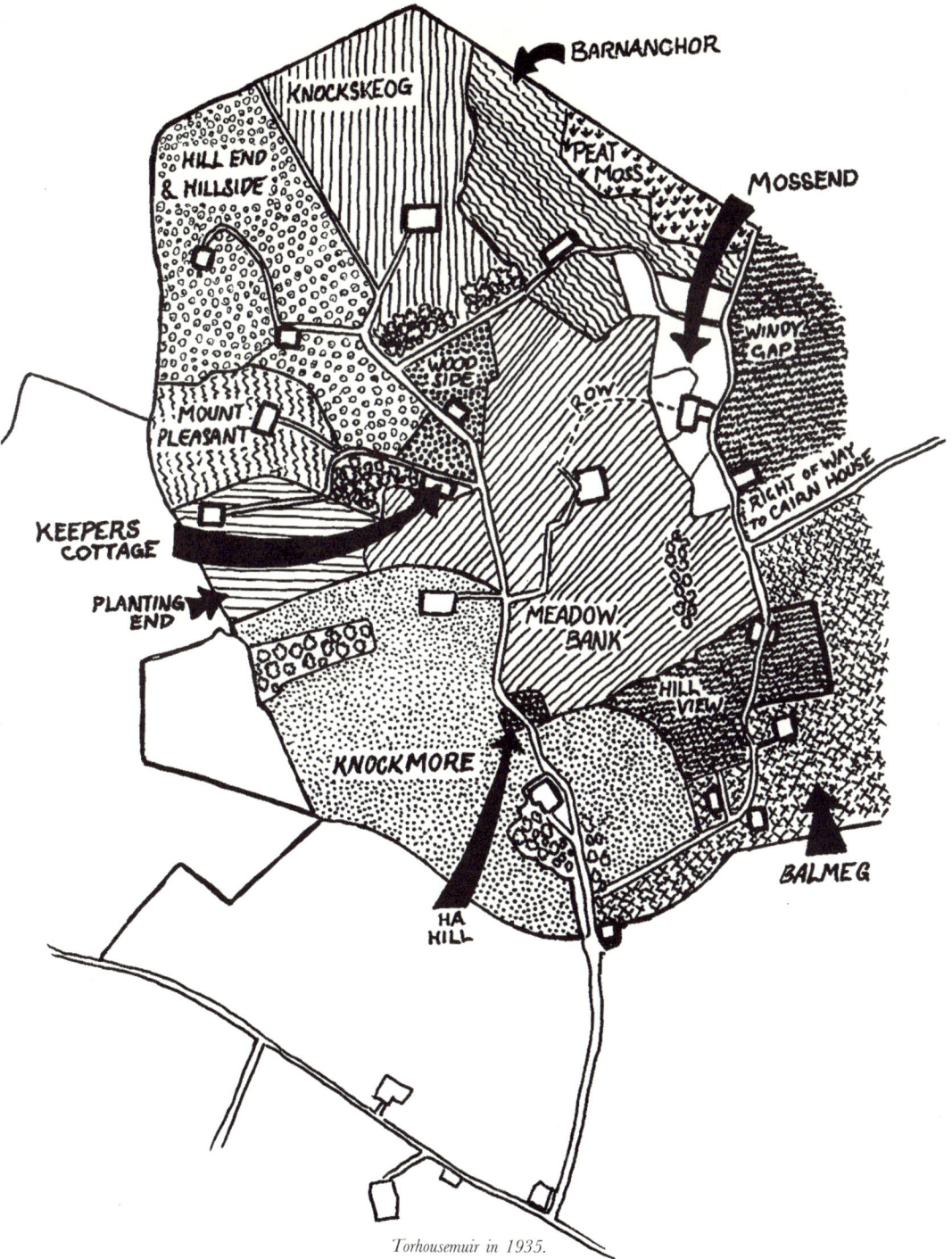

KNOCKSKEOG

BARNANCHOR

HILL END
& HILLSIDE

PEAT
MOSS

MOSSEND

WINDY
GAP

WOOD
SIDE

ROW

MOUNT
PLEASANT

RIGHT OF WAY
TO CAIRN HOUSE

KEEPERS
COTTAGE

PLANTING
END

MEADOW
BANK

HILL
VIEW

KNOCKMORE

BALMEG

HA
HILL

Torhousemuir in 1935.

5

The crofts are arranged more or less in a horseshoe, with the bigger farms of Knockmore and Meadowbank in the centre. Mossend was the furthest croft on the eastern side, and was connected by a right of way to Barnanchor. The landscape around is dotted with drumlins – islands of boulder clay – and in between is some arable land and rough grazing. With the exception of one or two plantings, the country at the time we moved there was generally open.

I began the story of my life at Mossend on the road in, and since Mossend was at the end of that two-mile stretch of unmetalled track, which meandered between all the crofts of Torhousemuir, the road was central to our existence. It was a cart track, lined by hedges and dykes: that, of course, is different from the tracks left by cars or tractors, since there were two grooves worn by the carts, and a path in the middle worn by the horses. We ran down it for school, carted milk to the road end for the creamery lorry, ran errands for our neighbours, cycled down it in our Saturday night finery, or crept along it when poaching; my mother, on the other hand, walked it on dark winter afternoons with a five-stone sack of flour on her back, bought from the mobile grocery at the road-end. I came to know its every inch: the banks half-way up which were alive with rabbits; the great dip which stopped the steam-thresher from going any further; the bridge beyond which you could not expect to find fish because beyond it the water ran directly off the moss and was too peaty for the trout; the part which was straight and where the footing was hard and rocky, and welcome when you were transporting milk or groceries; and the bend where the *fairy-tatties* grew. There were the woods where I was taught to make a woodsman's cross and learnt a lesson into the bargain: after cutting the cross on a stump with an axe, when I was left to take off the blindfold put on me by the woodcutters, I found that I had cut a cross right through my cloth cap.

Walking it, you have to imagine it as it was then: narrower than it is now, because the carts were smaller than today's tractors, and with the hedges nearly meeting over the top of you. In fact, the dense growth meant that, when the wartime black-out came, this was a place where you could risk showing a bit more light from your dipped bicycle lamp. You started at the road-end. This was not just the end of the road, but an important place: it was the point where, every morning, the carts loaded with milk-churns from the crofts would gather, waiting for the lorry from Bladnoch Creamery, and where news would be exchanged between neighbours. It was also one of three or four unofficial meeting places in the parish for the unmarried men, who would gather there on summer evenings. Somehow, everyone would know that on one night it would be the Malzie road-end, and on another the Ring bridge, or Torhousemuir; it was not organised, and yet everyone knew the rules and the times. On those nights prior to the outbreak of war, men would gather, sitting on the bank with bicycles strewn all over the roadsides, while others sat supporting themselves on their bikes, one foot on the pedal and one on the ground. The field just up the road, at Torhousemill, was also Torhousemuir's football pitch, and summer league games were played there, against Whauphill or The Grange, by men and boys who wore tackety boots instead of football boots.

The crofts were all between 20 and 40 acres, and we were all tenants of the Torhousemuir estate. Most of the crofts were on poor ground: it was either hard and poor land, or soft and mossy. Tenants, like my father, who had smaller crofts, would help on the larger ones, especially during hay and harvest. Our rents were collected every six months, and the leases were for a ten-year period, with a break at five. Many people left after five years and slipped back into employment, and I suppose this was a proof that conditions were poor. Some of the younger ones found that they could make better money by going back into employment. But of those who stayed – the McGhies, the Gemmils, McCallies and ourselves – all moved into larger farms and eventually into ownership of farms.

Our landlords lived at Torhousemuir House. When we first moved there, the owner was Lord John Fitzroy, who was somehow related to the Dukes of Grafton. He was much disliked for his drunken and arrogant attitude. We were taught to take off our caps to him, but we would have done anything to avoid meeting him, taking to the fields at his approach, and we secretly took out our resentment on his son, whom we punished mercilessly in fights. This son was eventually killed in a private aeroplane crash, and I heard that Lord John died in a padded cell. On one occasion, I can recall my father attending the traditional annual rent-day meal at the big house, but I think that this tradition was dying out.

As you go on up the road, the first croft was Balmeg, occupied by Mr. and Mrs. Tom Birch, a couple in their early sixties, with their one daughter, Jessie. The three of them managed the croft alone: most crofts were run like this by family units, or couples on their own. There was a dwelling house with a recently built brick extension to house the dairy and back-kitchen. The stackyard had room for eight stacks - usually six of corn and two of hay. Their fields were on the right of the road, and included a lot of rough ground amongst the arable land. They had seven cows and two horses, and also a corrugated iron pig-sty. This led to the unfortunate accident with the two sows, which were both electrocuted during a storm, when lightning hit the wire fence nearby. When smoke was also seen coming from their henhouse, the Birches must have thought another accident had befallen their livestock, but it turned out to be Jessie, who rather regularly checked on the hens in order to smoke a forbidden cigarette. Jessie, who was in her twenties, was never allowed out on her own and had few enough pleasures. Eventually, she escaped Torhousemuir, when she met and married a man who was working for Jones of Larbert, a company carrying out timber-cutting on the estate. Her parents themselves probably left the estate only once or twice a year. On one of these occasions, I travelled with them in a gig to Wigtown Show in 1936; I remember that the ponies were stabled at George Wallace's coal yard during the show day. I can also remember Tom Birch attending a funeral, clad in his bowler hat and funeral suit, whose colour ranged from the bottom of the jacket, where it was black, to a coppery green on the shoulders, where it had faded with age. The Birches had left the estate by 1940.

On either side of the road at this point were two ruins – old dwelling houses which had fallen into disrepair. These had previously been occupied as crofts in their own right, but were now abandoned, and their land taken over by the neighbouring crofts.

Mr. and Mrs. Adair lived in Hillview, so called because it looked towards Ha' Hill, the hill where later my father would take his tour of duty with the local wartime Defence Volunteers, manning a look-out post. When I knew them, they were in their sixties. Hillview had taken over the lands belonging to Windy Gap, the croft further up on the right hand side: this consisted of one arable field, with the rest in rough grazing. They had a newly built byre, which could house seven cows; they also had a new kitchen and dairy which had been added to the dwelling house. Parallel to the house, there was the old byre and barn, with a cart shed with three bays, and a stable. The Adairs kept goats, which ran off in a thunderstorm and were urgently searched for over a whole day. They were eventually found in a sorry state, because they had been gorging themselves on plums and suffered the consequences for days afterwards. The Adairs had a phaeton, a gig and two bicycles; gigs were already going out of fashion, and were only used to fetch people from Wigtown station. It always puzzled me that when Mrs. Adair was at a loss to answer any question she would say: "I was born too early in the century", though she must have been in her sixties in the 1930's. It was a common practice to identify people by their sayings and there was plenty of scope for people at Torhousemuir to develop eccentricities.

Approaching Windy Gap, there was, on the right, a right-of-way to Cairnhouse. In the 1930's, it was already overgrown, but you could still glimpse a roadway leading to another ruined cottage. You could more or less follow the route by zigzagging through the whins; eventually, it joined up with the Old Edinburgh Road. Windy Gap, the last croft before our own, was rented by Duncan MacGregor. It consisted of a dwelling house of two rooms and a living room, an old barn and a stable, all built in one long row in the usual crofting style. Part of the building was roofed with corrugated sheets. In front of the house was a kitchen garden. Duncan's parents had farmed the upland farm of The Buchan, but on their death, he had moved to the Machars and eventually to Hillend on Torhousemuir. His last move was to Windy Gap and when I remember him, he was probably in his fifties or sixties. He was something of a recluse, and I never knew him to go further than a travelling grocer's van at the road-end. He had the reputation of being lazy, but since he had served in World War I, it might have been the effects of his war experience. Certainly, he only worked in the summers, at turnip-thinning, hay and harvest on other crofts; as I said, the land belonging to his croft had been acquired by the Adairs. He milked the cows at the croft of Hillend, but if there was a threat of rain, he would simply say: " It's gey wet-lookin'" and refuse to take the churns to meet the creamery lorry. Like us, he also cut peats from the moss, but having no cart, he had the heavy work of carrying them home in a sack. He always seemed to have enough money for food and tobacco, and my mother would cook him lunch on a Sunday, which we took to his cottage. It was full of peat-smoke, and his skin was always brown with it. He stayed at Windy Gap until he was no longer able at some point in the 1950's, when he was probably in his seventies, and was then taken into care.

Our own croft was the outermost holding, and beyond was the *Moss Road*, leading over the moor towards the Woods of Auchleand farm, which had to be reached via the Mains of Penninghame. Woods Farm was legendary even then for the dirt and meanness of its housekeeping, and during the war was one of the few households to be refused as suitable housing for evacuees on an inspection visit by the local doctor. I can remember being chased off the moorland by Andy and Davy Ross, the father and son, when we had been in search of whaup and gull eggs.

Our croft, Mossend, had been occupied by Mr. and Mrs. Love, until John Love had died suddenly in 1935. It was connected by a walking right of way to Meadowbank, one of the larger farms. This was the way we went to meet the grocer who would travel as far as Meadowbank, and it was also used by the postman.

Beyond our own croft, and west across a marshy ground, there was the croft called Barnanchor, which was farmed by the Menzies family, who moved out before 1940 when the parents of the family reached retiring age. Then it was taken over by the Edmonds family, who had come from the Newton Stewart area, at Barnkirk. Barnanchor, like Mossend, had the remains of a large horse-driven mill to the rear, but these were disused by the time we arrived in the 1930's. Mrs. Edmonds was overweight, and could barely move about; she never left the croft – until eventually she was carried out of it. Whenever she met us during the 1940's, she would always ask: "Is there any war news, boys?" Now I can appreciate better why she was asking, though it did not occur to us at the time that there was a reason for the endless repetition: she had sons-in-law in the services, and one at least was a prisoner of war in Singapore. Jimmy, her eldest son, stayed at home and went to work at the munitions factory, which was set up at Carsegowan, over the moor from Torhousemuir. It operated on a three-shift basis, night and day, and for five years, Jimmy walked over the moor to the factory, carrying a candle in a jam-jar to give him some light as he crossed the moss. On a windy night, it must have blown out before he had gone ten yards. He must have had quite some strength, because I remember his going this way across the moor carrying two cart-shafts tied over his shoulders with a rope, to take to the smith at Causewayend. Aaron, her younger son, was my own age, and came to be a friend.

The Edmonds had seven cows in the byre and one horse; they did not carry out much arable farming. They were the only other family apart from us to use the peat-moss for winter fuel. There was a byre, stable, barn and another outhouse, with a vegetable garden in front of the house. Rachel Edmonds, the daughter of the house, carried out most of the work on the croft, and would take the spring cart down the road in the mornings to meet the milk-wagon. Rachel obtained clothes from her sisters in the cities, and was often to be seen wearing fashionable hats, riding her old rickety bicycle down Torhousemuir's bumpy road. Rachel's story was one of Torhousemuir's romances, since she eventually married Willie Lindsay of Meadowbank, who was 25 years her senior. When the news broke, we could scarcely believe that she would marry someone so old, and a foot shorter than she was. Besides, we knew that Willie, who was fat, had to have his trousers cut off at the knee by my mother, who was a handy seamstress. The Lindsay family were more than surprised, however, since Willie's mother and two sisters vowed to keep his bride out of Meadowbank, so that poor Rachel, even after marriage, had to stay at Barnanchor. Rachel stayed on at Barnanchor after her mother and brother died, and, only after her mother-in-law died, she was occasionally allowed to stay at Meadowbank. Eventually, she moved to Newton Stewart to live with her husband and sister-in-law.

Torhousemuir crofters at harvest-time.

Knockskeog was occupied by the Horners, a couple who were in their sixties when we arrived on the estate, and who had a long family history on the estate. They had about seven cows, a barn, hayshed, and stable. Two of their sons eventually joined them, the youngest of whom, Drew, took over Knockskeog in the late 1940's. He had been at Mount Pleasant, also on the estate, prior to this, and later retired to the gamekeeper's cottage, leaving Knockskeog to his son. This is the only Torhousemuir property to remain occupied by the same family until today.

Hillend, one of the smaller crofts, was occupied, when we arrived in 1935, by Mr. and Mrs. Dan Love. It was then taken over by Mr. and Mrs. McClymont who had three grown up sons, all of whom left for the armed forces early in the war. Their parents did not stay long after this and the call-up affected many of the crofts in this way. Hillend had already taken over the land belonging to another adjacent croft, Hillside, where the dwelling house remained vacant. I can only just about remember Hillside being occupied as an independent croft, at the beginning of our time at Torhousemuir. When the McClymonts left, the entire holding was rented by the Campbells of Knockmore.

Mount Pleasant was farmed by Mr. and Mrs. John McGhie, who eventually moved to a large farm on the Mochrum Park estate. Mount

Pleasant eventually became fused with the lands of Planting End, and was farmed by a Mr. McCallie. Planting End, which also came to be called Woodside, seems always to have been occupied by people who stayed only a short while. Woodside was occupied by a retired blacksmith called Lancelot McKenna, who lived there with two daughters. Both these small crofts had to be reached by cart tracks, off the main thoroughfare at Torhousemuir. McKenna kept only a few cows to produce for his own consumption. Once he had left, the croft was taken over by the estate.

Circling back down the road, you came to the Gill's cottage, or the Keeper's Cottage, which had no land of its own. Andy Gill was gamekeeper and had five daughters and one son. Two of the girls were in domestic service at Torhousemuir House. Andy, the youngest son, never came back to Torhousemuir after he went into the Army, except on holiday, and he married an English girl.

Knockmore and Meadowbank consisted of bigger acreages altogether; in fact, they were farms, not crofts. As I said, Knockmore eventually came to include Hillend and Hillside. It was farmed by the Campbells, who had been at Torhousemuir for several generations. The farm had been inherited from their parents, and the spinster and her bachelor brother farmed the holding together, after Jim had come back from the First World War. They were well-connected people, different from the other crofters. It was the Campbells who had the car for the school run, one of three at Torhousemuir in the thirties. Miss Campbell was our driver, but was always very nervous, and – like everyone else on the road at the time – had never passed a test. I can remember hurtling down Kirvennie hill at 60 miles per hour and wondering if she was in a hurry to make the harvest tea. The Campbells had as many as thirty cows in a new byre, four horses, and a new hay shed adjacent to the byre. They had a dairy and other outbuildings, a windmill pump and a gravity-fed system linked to the farm.

The Lindsays farmed Meadowbank, after moving there from the Kirkmadrine area, where they had been dairy folk, in the late twenties or early thirties. The farmhouse there had the grandeur of both a front and a back door. The stable was adjacent to the house, and the byre held twenty cows. There was a small byre for younger cattle, and a dairy. The farm extended over about seventy acres, which included two big hills and the rougher land below. There was also a large barn. The mother ruled the household with a rod of iron and it was she who said that Rachel Edmonds, who married her son Willie, would "never be in Meadowbank while she lived". She had had four children, one of whom had died before they came; the surviving son, Peter, by then in his mid fifties, had been gassed or shell-shocked in World War I, stuttered and had a habit of repeating his questions constantly. He was given the job of carting milk from Meadowbank all the way to Bladnoch, despite the fact that a wagon came to the road-end, just to keep him occupied. When running late, he could be seen gulping his breakfast while driving the cart down the road. He was often in trouble because of his slowness and my father and my brother Jimmy were witnesses to one incident where he was severely punished. Peter had been left in charge of carting turnips; when carts had to go uphill with a heavy load, it was essential to fill the cart only half full, or else the horse would *hang itself*. When this happens, the horse's collar cuts off its air supply and it drops, usually on its side. When it revives, it struggles up, lifting its head and choking itself again. If such a disaster occurred, the cure was to sit on the horse's head to stop it from harming itself. It then had to be unyoked from the cart and allowed to stand up, while half the load was removed. You also had to re-yoke the horse as soon as possible to enable it to get its confidence back as soon as possible after the fright. My father and Jimmy saw from the field they were working in that Peter had got into trouble with his horse and dropped all they were doing to run and help. By the time they arrived, Peter's mother was on the scene, gathering up the spilled turnips in her apron, shying turnips at Peter's head and cursing him, while shouting: "You should have been drowned when you were a pup!" Jimmy remembers dodging the turnips, and no doubt Peter remembered the incident too, because he was given no dinner and not allowed inside the house for the night.

Willie had a Ford car, and was the only one ever to leave the farm, attend the market and – in fact – the only one to get married. He was still under the thumb of his mother, who controlled strictly the number of cigarettes he was allowed per day. There was a constant battle of wits between her and Willie as to where they had been hidden; he seemed to develop a cunning for finding them out, even if they were hidden in a bag of flour. Willie's courtship had probably begun when he and Rachel Edmonds went to meet the grocer's van at the end of the road, and we used to see them sitting on a wall on Rachel's way home. Even after the old mother's death, one of Willie's sisters conspired to keep Rachel out of Meadowbank, and eventually, Rachel was left on her own, looking after her own mother, while Willie and his sister Jessie moved to Newton Stewart. It is hard to explain the attraction Willie might have had for a young girl. For example, there was a strange habit of the Lindsay men which I remember well: their underwear, which was worn until it virtually fell off in tatters, was left hanging over the stalls in the stable; after many years, there was a huge and unsavoury accumulation.

Kate and Jessie, the two daughters, were both over-weight, and, perhaps as a result, had the equally unusual habit, which fascinated me as a child, of putting on their pink laced-up corsets over their dresses. Kate often wore men's vests and long-johns and the whole outfit was crowned with a dust-cap. A degree of isolation was possible then which it is quite impossible to imagine now: the Lindsays might have gone for weeks seeing no one but the grocer and the neighbours, who did not count. As a result, eccentricities grew and no one saw any reason to correct them. At hay and harvest, when the two sisters sat down to tea in the field, they had to roll over onto their stomachs when they were finished in order to lever themselves up from the ground. Their clothes might have been dirty and rather odd, but their kitchen was spick and span, and they were kind-hearted. If their newspapers had missed collection by the post, I would run them up when I was dropped off by the school car, and receive a penny and a huge soda scone, with blackcurrant jam. They had an orchard and soft-fruit bushes, kept hens and ducks, a couple of pigs, and over twenty cows.

Ha' Hill was occupied by Miss Ella Laird and her brother, Willie, who was slightly simple. He used to ask us, though we were just children, on seeing a young woman pass and rubbing his hands together and repeating himself: "D'ye think she'd court, boys?" They had a private income, possibly inherited from relatives in the drapery trade in London. Other relatives farmed at Redbrae and Torhousekie. At Torhousemuir, they had only one small field and a garden. It was one of the strange Torhousemuir rituals that Miss Laird unfailingly entertained Dr. Watson, the local doctor, to tea for the whole of every Tuesday afternoon.

The Gardener's Cottage went with the job of gardener to the big house. The last gardener to occupy it went by the name of Torbet; his predecessors were a Mr. Smithers and Mr. Cruikshanks. Then the Wilsons, who came from Belfast, rented the cottage and settled there in 1940; Mr. Wilson worked at Carsegowan and Baldoon.

THE CROFT

MOSSEND was a croft of twenty-five acres of land, mainly arable, but with about one and a half acres of heather. My father was in his fifties when he took it over, and the rent was ten pounds per year. I am sure that none of it was ever borrowed, because my father avoided debt as something to be feared. Throughout his long life, if he had no money to buy a thing, he simply did not get it. To stock the croft, we had to buy a horse, cattle, implements, dung and crops from the previous tenant, which were valued by an assessor agreed by the incoming and outgoing tenants. When all the expenses were paid, my father had just sixpence in the Bank, but his account was in credit.

Mossend was approached over a gentle rise in the road so that only when you were close could you see the whole house. Before you saw it, however, you could smell the peat reek from the chimney, since the peat fire burnt all year round, because it was the sole cooking source and was burning low even when everyone was away in the fields. On your approach, you could hear the cackle of the chickens about the door which would come and meet you when you approached. You could see the roof over the hedge, and then rounding the edge of the dyke, you were standing on the beaten earth of the yard, with the peat stack on your right, the house in front, and the milk house to the left. Between the dairy and the end of the byre, there was the wall beyond which the midden was heaped. The house was built of stone, consisting of one long structure, including the barn and byre beyond. It was whitewashed, and the doors were painted. The roof of the house was slated, but the barn and byre were covered with corrugated sheeting. The house consisted of a main room and two bedrooms, one of which was brick-built and had been added on to the old stone structure at a period when the estate was being improved. At this time, elsewhere on the estate, new byres had been built in brick with asbestos roofing.

You entered the house through a tiled lobby, which had tongue in groove wainscoting on the right. We painted it after arriving at the croft, after removing fourteen layers of wallpaper. Just beyond, and sandwiched in between the living room and the bedroom was a very small scullery, where foodstuffs were kept. My brother Willie can remember being kept awake at night in the adjacent room by an old rooster which had been sacrificed for the pot, but resolutely refused to die and kept up a cackle all night. As well as functioning as a larder, the scullery also housed the bucket which was used as a chamber pot at night: no one had any sense at this time that this might have been unhygienic. The lobby and the kitchen were floored with white clay tiles, which were worn hollow in many places. Eventually, we covered over the tiles with linoleum, which also sank into the hollows.

The main room served us as dining room, sitting room, a place to gather at night and an extra bedroom for my sister, Lena, who was still at school when we arrived at Torhousemuir. The walls were plastered straight on to the stone, and in our time were wall-papered with a print of ferns in orange and brown. There was an open fire in it, with a mantelpiece about six foot above the floor. Socks were permanently hung to dry along a line suspended from the ledge; if the day had been wet, there might also have been two chairs to support the sheets drying in front of the fire. The fire itself was about three foot wide, which my father reduced to eighteen inches, to form two hobs on either side. The hobs were brick, with metal covers my father made out of tin, to conduct the heat. The fireplace had a deep fender, the lower part being black-leaded, the upper part polished to a shine with steel wool. The fire was about 2 foot six inches above floor level, and the ashes fell onto the floor below. It was fuelled solely by the peat we cut ourselves on the moss. In fact, the sole heating in the house was from the peat fire in the main room, which warmed your front, but left your back cold. One trick was to put your coat over the back of your chair, to cut out the draughts from behind. By the fireside and also in the bedrooms, we had home-made pinned rag-rugs.

The lighting was provided by oil lamps – a double burner in the kitchen and single burners for the bedroom. The double burner could not be turned up too high, or else it smoked the funnel, so that the light was considerably less than people are now used to. It was a daily chore to clean the globes and replace the paraffin before nightfall. For the byre, a Tilley lamp was used, which consumed pressurised paraffin, and gave off both heat and light. By the War, we had an Aladdin lamp in the main room, which had a circular wick and mantle, and gave off a bright light, which was said to be the equivalent of 100 candle power.

By the door, there was the meal ark, all important because it held the oatmeal for the porridge. It was about four foot high at the back, had a sloped lid and drawers, in which baking powder and soda were kept, and was about 20 inches wide. Next to it, along the wall, was a bench or form, and the table, with the bath full of peat beneath it, and kitchen chairs ranged along the other side. On the wall opposite, there was also a dresser and a cupboard in the wall. By the pantry, there was also single bed for my sister Lena, and by the fire were two easy chairs in Rexine for my parents.

Privacy as we now understand it was non existent in large families of this time and in this we were no different from anyone else we knew. The bedroom to the right housed four of us: my parents in one double bed, and my brother Willie and I in another. It was an iron bed, which could sleep three if visitors came. I was often given the hard place in the middle, being the youngest, so I preferred to be relegated to a *cauff* mattress on the floor. The cauff bed was warm, and fortunately, the bedroom floors were wooden and less hard than the kitchen and hall. Frequent visitors were my three cousins from Ayr, who cycled down, and expanded the household to ten, or perhaps twelve, at weekends. The other bedroom, to the left of the front door, was occupied eventually by my sister, Jean, who returned to us with her daughter, when her husband, a gardener at Barcaple, near Ringford in the Stewartry of Kirkcudbright, joined the medical corps. Jean left us eventually in 1944. The proximity of the barn meant that at night, in any of the bedrooms, we could hear the mice jumping the rafters above our heads.

Immediately outside the front door was the hand pump, against the wall of the barn. At first, the pump, which drew on a well to the rear of the croft, was simply outdoors, and bath nights were held in the kitchen: water had to be boiled in large pots on the fire. Eventually, my father contacted the factor, John Black, who was a Banker in Wigtown, and he agreed to purchase corrugated sheets for the building of a wash-room. My father and my brother built the corrugated iron shed over the pump, and added a cast iron boiler with a wooden lid, which would heat the water for washing and for the baths we took at weekends: in those days, there was no such thing as daily hygiene. For both washing and bathing, there was the galvanised iron bath, which was hung up on nails on the wall of the wash house, and there was a rubbing board for scrubbing clothes. The boiler was heated by logs which we cut on wet days, or which my brothers would cut at weekends; its chimney was vented through the roof. It had a dual purpose, since it was also used for boiling up the pig food. Buckets would also be drawn off the pump each day and kept in the lobby for use in the kitchen or scullery. Despite the added refinement of the shelter over the pump, it was still extraordinarily draughty in winter. We would economise on well-water in summer, by drawing on the peaty water of the burn which ran out of the moss, but even then, the well still ran dry by May. Some crofts were lucky enough to have dipping wells for drinking water, but there was none at Mossend.

There was a dry privy or closet at the top of the garden: to reach it you had to cross the yard, go past the milk house and at about fifty yards from

Mossend.

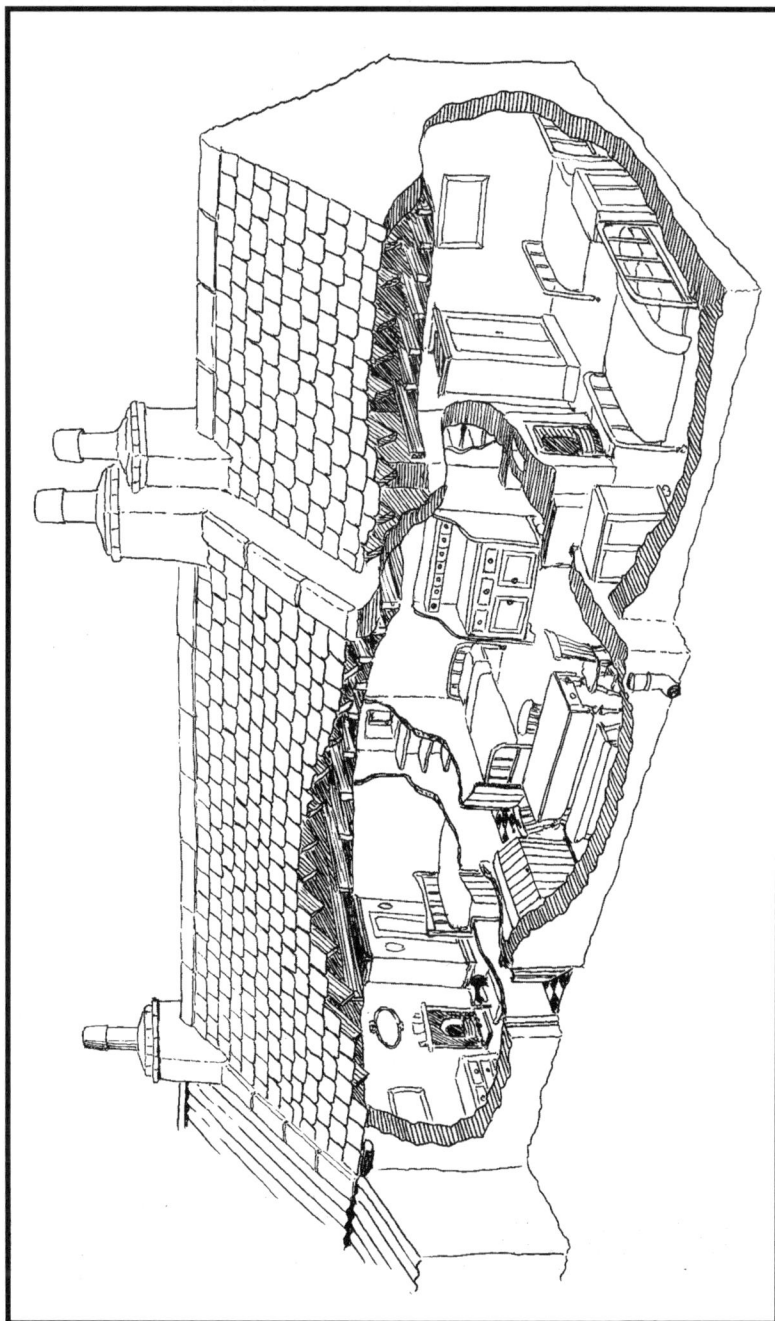

Cutaway reconstruction, Mossend.

the house, there was a wooden building, with a seat and a bucket below. This was emptied onto the midden in the farm yard every four or five days, and consequently could be scented downwind at least ten yards away. No other croft had anything more luxurious, except for Knockmore, which may have had a flushing toilet. My mother thriftily saved tissue paper which separated the sterile milk pads used in the dairy, and hung them on a hook in the closet. In fact, most of the men of the household did not use the closet, and simply found their own way among the whins at the back of the house.

Beyond the barn was the byre, where the five cows were milked. Most of the crofters had so few cows that they were known by their pet names, and some even chalked the cows' names above their stalls. It was the time when concrete stalls were coming in, replacing the old wooden stalls, because sanitary regulations and inspections were beginning to apply. Almost opposite the byre was the dairy, where the milk was treated and sieved, before going in the churns to the creamery.

At the end of the building was the stable with a corrugated iron roof, where the two horses were kept. There was a heavy pony, Jimmy (after the man who sold it to us) and an Irish half-bred pony, Billy, which had been bought at the Newton Stewart horse-sale, which was held every October. Of the two, the pony was more temperamental, and could bite. It was eventually considered too light when we moved to Torhousekie in 1945, and was sold. These same two horses performed all the jobs on the croft - ploughing, reaping, and carting – and they were the engine without which the whole croft economy would have ground to a halt.

The fact that the byre and stables were so close to the house meant that we lived with the sight, sound and smell of the animals day in and day out. In summer, the croft door was simply left open all day until nightfall. You could hear the bellowing of the cattle waiting to be fed at 5 o'clock, or the sound of cattle getting up or lying down in the byre, and contented grunts of the horses when well-fed at night. The byres were often warmer and more welcoming in winter than the house. If the cattle were let out in winter for a drink at the burn, they would gather at the gate waiting to be let in. In winter there would be the steam coming off working horses in the fields, and steam from the dung on the midden. You lived day in, day out with the animals' habits and depended on their intelligence and training: the horses would *nicher* as you went in, giving you a greeting at the start of a day; you would know if they were upset, or would dislike the weather.

There was a new cart shed, which we erected ourselves, next to the dairy and at right-angles to the house. It replaced the old cart and gig sheds near the moss road, which in our time became the peat store. We kept two carts: a spring cart for light work, such as carting the milk to the main road, and a box cart for heavy farm work. At one time we had a gig, but these were becoming outdated. I can only remember going as far as Wigtown in a gig, collecting my uncles from the Whithorn bus. The motor-hire companies, which did the work of a taxi service, really spelt the end of the gig.

Facing the stackyard was the henhouse and the pigsty. We only kept two sows and eventually we got a boar prior to the War, at which point they had to be registered. Before we acquired the boar, we used to have to take the sows in an open crate on the back of a cart some eight miles. She would be left there for about a week to be served before returning home. The hens were my mother's preserve: she fed them and made sure they were closed in at night. She had an easy knack of breaking their necks quickly with one hand, and she used to hang them upside down in the scullery. Nothing was wasted, and their feathers were, of course, used for stuffing pillows. Later on, during the war, we used to buy day-old chicks in boxes of twenty-four. We put them in a home-made brooder, which consisted of a run about three feet long, and a compartment with a light in it and a cloth. The chicks got underneath it, as if it were their mother, since it gave off heat. The cocks among them would go into the pot, while the chickens would be kept for egg-laying.

The stackyard was only used for hay-stacks, since the corn-stacks had to be built down the road, because it was impossible for the threshing mill to reach the steading, owing to the hollow in the road at the burn. From June until the end of March the following year, there was the familiar sight of the domed hay-stacks in the yard. At first, early in the season, there would have been four, and the number gradually dwindled as we used the hay for fodder. There was also a fairly large garden west of the stackyard, and at right angles to the steading, where there were soft fruit bushes at the foot and vegetables, which my father would put in every year, grown for the table. It was surrounded by a hedge. The washing line for the clothes was stretched at the bottom of the garden; on a Monday it would be burdened with washing for six or seven, including the grown up sons who worked away and brought their weekly laundry home.

The gates on the crofts were not metal gates as you see now, with spring-loaded bolts: they were quite often hand made with two bars of straight-ish branches and were seldom, if ever, hung. Often, they would be removed to another field, if they were needed there. Another way of barring a hole in the hedge, or *slap*, was simply to cut two large thorn bushes, pull them towards each other into the gap and tie them in place.

Daily routine

T HE croft economy worked because cash expenditure was kept to a minimum, and the most was made of the labour of the crofters themselves. Usually, the crofts depended on the constant labour of both husband and wife; most of these were elderly when we arrived at Mossend. If possible, crofters would also contract their labour out to other farmers at slack times, or if they had completed their own work more quickly on the smaller acreages. This meant that anything which could be made on the croft, mended, or made to do, without spending cash for repair or for new items, had to be done. It meant that my parents had to be very resourceful as well as very hard working people, and that each member of the family had to pull his weight. Waste and miscalculation, failure to plan for the winter ahead, could have brought ruin upon the entire family. There could be no letting up in the relentless routine of work. The new horse-drawn machines which were coming into use before the War were often too expensive to be justified on a croft: we had to consider the acreage and the use the machine would be likely to get in a season. We therefore often kept to older methods, with less expensive machinery which required greater labour, and, if necessary, we borrowed or hired equipment from larger farms.

There was a daily rhythm to be followed at Mossend, slightly different for the different seasons, but always unvarying in its essentials. Mother was always first up, and, after building up the fire and putting the kettle and porridge pot half on, she would go straight to the byre to milk the cows. By this time, father would be up and would *feed and fodder* the two horses and the younger cattle, which were housed in a lean-to towards the foot of the stackyard. He would help to finish off the milking, but my mother could always milk two cows to his one. He would then take the milk to the dairy, put it through the cooler, and into the churns, ready to meet the creamery run.

The two boys who were still at home, but working on neighbouring farms, would find the kettle boiling and the porridge warm enough to eat. They would also fry a breakfast of eggs, potato, soda or oatmeal scone and bacon in lard or dripping, and leave for work. Father and mother would return after milking and eat porridge and a fried breakfast. The horses would then be yoked to the cart, and the milk taken to Torhousemuir road-end to wait for the creamery lorry. Probably, the cart would return at about 9.30am. After breakfast, the oatmeal would immediately begin soaking for the porridge next day; only rarely did we eat yesterday's *back-het* porridge, and mostly the remnants were fed to the hens.

Father would go off to the fields, and might take his lunch there, if he were busy with the ploughing. Even in winter, there was fodder to bring into the byre, corn to bruise, turnips to cut, or treacle mash to prepare for both horses and cows. My mother would bake, prepare food, and look after her hens. At night, the routine of the morning was repeated, because it was time for the horses to be bedded down and for the cows to be milked again. Usually, even when I began work on the croft, I did not milk, but in harvest-time there was no choice: we all mucked in at that time.

When we were school-children, we had a series of tasks to carry out when we got home from school. According to the season, we fed the young calves with a bucket of milk, fed the hens and gathered the eggs, helped with turnip-hoeing, and hay and harvest. In winter, when the cattle were inside, we would carry buckets of water to them, and feed them with three or four turnips each, and a bunch of straw between two animals. Then we had to get the barrow, and clean the dung from behind them. At weekends, we would curry and brush them. It was only once all the cattle and horses had been seen to that we would have our own supper. These were not jobs which we resented, and often we would compete to see who could complete the job well, or best imitate the adults.

By 6.30pm, especially in winter, everyone would settle down around the fire: we might read or do schoolwork. Sometimes, we would play draughts or dominoes, unless the table were required by my mother, who had to catch up on baking at night if there had been urgent outdoor work to do during the day. If she were not cooking, she would sit and knit, sew and mend, or make a rag rug for the fireside or bedside. My father would sit and smoke, or lean over the stackyard wall. He would spit occasionally into the ashpan, creating a huge cloud of peat dust. My mother was always last in bed, and would move the porridge pot and kettle closer to the fire, wind up the clocks, set the alarm and put all the lights out.

In winter, we would be in bed by 9.30pm, but in summer, when we had double summer time, it stayed light until almost midnight and it was hard to sleep.

ROTATION OF CROPS

THE main activity of all the crofts was dairying, and all the raising of crops was dedicated to providing fodder, promoting milk yield and over-wintering the cattle. We grew oats, turnips, hay and straw, and these were in universal use for feed and bedding in the days before silage. We followed a rotation of crops on the fields which was the norm at the time. The croft had seven fields, all of which had individual names, according to their geography or condition: there was the *far* field, the *moss band*, the *far rashy* and the *near rashy* fields, the *wee* field, the *stackyard* field, and the *back o'the house* field. Those which were cultivated were done so according to a rigid rotation, although some might be halved, so that the two parts were being treated in different ways in the same year.

The basic rotation was as follows: corn (that is, oats) was planted the year after a grass crop. When the corn was a stubble field, one year later, it was used for a root-crop. The year after a root-crop, the ground was known as *red land* or *rid-land*, because there was no grass planted. This was ploughed and sown in oats, as the nursery crop to a grass crop, which formed a hay-field the following year. When the oats were first taken off, the grass was called *young* grass. Usually, this was grazed for a short period, and then manured and left to grow for hay in the following year.

The rotation also determined the sequence of tasks during the year, the way in which each field, at any particular stage of the cycle, had to be treated, and the tools which were used. I describe it, because these things have changed, following the introduction of more powerful machinery, different varieties which can grow more quickly, or at different seasons of the year. Beginning in November or December, the stubble fields would be ploughed. The sooner the stubble was ploughed and dung-ed, the better it would break down, and the more easily it would be worked for a root-crop. The next stage was to plough the *lea*, or grass, which might have been under grass for at least two or three years, or sometimes as much as 7 or 8 years, and which was to be sown with oats. The last ploughing of the year was of the red land, for sowing with corn and grass.

Early in the season each year, we would go out and burn the whins on the knowes, which had a tendency to take over, if left unchecked. For this, we used a home-made torch: we would tie a bunch of hessian sacking onto a piece of fencing wire, soak it in paraffin, and force it into a cocoa tin. When lit, it would smoulder at the end, but to make it flame, you would *birl* it around. This would give you a permanent source of fire for burning. The burnt stumps were very brittle and the following spring you could collect them up and use them for kindling. My mother would use them for a quick burning fire, which was good for baking pancakes.

Corn was sown in April; by May or June, the turnips were ready to be thinned, before the hay harvest in late June or July. The hay harvest was best secured and home before the oat harvest. After harvest, which might, in those days, be as late as October, there were the potatoes to lift, in

October, and then turnips to *shaw*, so as to have feed for the cattle. There was also the muck to put onto the stubble for the next root crop.

The rotation was the same everywhere, and the talk between neighbours always related to the progress on the farm and, as a matter of the greatest importance, the weather. Weather, after all, could make or break a croft. A bad season meant that corn might be lost, unless it could be rescued by scything, and a loss might mean having to buy in cattle fodder during the winter. People were used to predicting the weather from the sky and the condition of the clouds: they would look for the higher cloud, which they called *goat's hair*, for a sign of good weather, or check to see if the clouds were scudding fast. Probably as a result, there were many proverbs relating to the weather, such as "March stour is worth a guinea an ounce", because a dry spring would let you get on with the preparation of the ground and the sowing.

PLOUGHING

THE ploughman was a man to be respected in the farming and crofting community. What a farm required was a good ploughman, a good horseman, a good stack-builder and a good drill-maker. A ploughman was supposed to be capable of doing all these things. He could assess how much dung to put on a field and calculate the right size of pile; he could add up the quantity of seed corn per acre. All these things he knew and worked out naturally. On a farm, the second ploughman would be a less experienced man, hoping one day for promotion to ploughman, for the prestige and the extra money. The rate of pay was negotiated between the ploughman and the farmer. A shepherd might earn more than a ploughman, but he worked less hours and did not have the prestige. Ploughman and farm were judged on the straightness of his ploughing, as summed up by the critical observation of neighbours. On the ploughman's skill depended the success of the sowing and the crop: since, in these days before dibblers, the corn was sown broadcast, the seed-bed had to be well-prepared by the ploughman if the rows of corn were to grow straight and without waste.

The plough we used for ploughing the lea, or *ley*, was a swing plough, which meant that it had either no wheels or only one attached to the beam of the plough; for stubble, we used an Oliver 110a *prairie buster*. The type of cut taken in each case was different, because you could take a wider cut in the stubble which was being prepared for a root-crop, whereas a narrower furrow was required in the lea since it was to be used as a seed-bed for corn. In the case of the swing plough, it had a wheel which was used to regulate the depth of the cut; this wheel ran along the landward side, which is to say the unploughed side of the field.

The coulter cut the sod to a depth of about three to four inches, and was positioned three fingers' width behind the sock, which dug the sod ready to be turned by the mould-board. The coulter was suspended about two and a half or three inches above the sock, and had a slight angle to ensure that the grass was cut so as to be turned over completely. Plough socks, of course, were pieces which could wear, and when used in stony ground, they would wear more quickly; if necessary, a piece could eventually be welded on by the blacksmith. As they wore, the ploughman had to know how to adjust the angle of the plough, so as to force it to dig more deeply, while, if the ploughshare were new, the angle at which it was held had to be lowered, to stop it from going in too deeply.

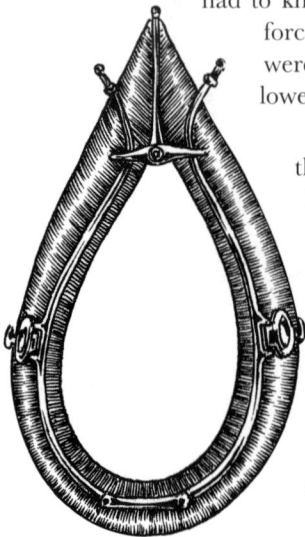

The horses wore a collar and *hems* for ploughing; the collar protected the horse's shoulders, and the hems were attached to the collar. On the hems was a hook for the draught chains, which went along the horses' sides and were attached to the *back-raip* or back-band. The chains were connected to the swingle-trees, and the *traip* or main tree. The ploughman controlled his horses by means of reins, which were held along with the handles of the plough, or *stilts*. There would be a mouth-cord connecting the two horses, so that pulling on the one also controlled the other. The mouth-cord could be slackened or loosened, according to the job in hand. There were also tie-backs from the mouth of each horse to the hems of the other, which crossed over between them and enabled the ploughman to keep them working in unison.

If the plough was set up correctly, a simple movement of the hand back along the handles would tighten the rein and guide the horse. The correct stance for a ploughman was deep into the stilts, so that your arms were straight down, and a slight pressure up or down, or a leaning to one side, would control the plough. That sort of intimacy with the plough meant, of course, that should you hit a stone, the plough could be wrenched from your grip and hit you hard in the ribs. A good ploughman had a complete understanding of his plough and its relation to his horses, of his horses' mood, and the nature of the soil he was ploughing. All this accumulation of expertise enabled as precise a treatment of the field as if it had been measured by a machine.

Balance had to be achieved when yoking to the plough, and a correct proportion achieved between angle and depth, but if you were holding a plough that was properly set, it was not difficult or heavy work. My father would let me have a *haud at the pleugh* when I was still a very small child, while he walked beside holding the reins, and turned them for me at the end of the furrow. When I was a little older, I was allowed to take charge of both plough and horses, and would take them once up the field and back down again, while he sat at his tea-break. The single length of the field – down it and back up to the start – was called a *booting*. It was a gradual progression in training which we were quite unaware of, so that by the age of 13, we probably had all the skills of a ploughman. We were keen to have the knowledge, since ploughing was an achievement which was admired, and we saw it as something to aim for. We also had to learn the knots to tie: when tying the reins, a bow-line hitch, known as a *ploughman's knot*, was used, since it could be easily loosened, as it had to be: if it rained the knot would tighten and become impossible to undo.

I probably first ploughed a full day myself at the age of nearly fourteen, when I had just left school. I ploughed a 10.30am to 3 o'clock day, for about two weeks. The routine when we were ploughing began with the feeding and foddering of the horses at half past six in the morning. They would also be curried and brushed with a dandy-brush at this time. Their tails were plaited to keep them cool: the hair was divided into three sections, plaited with the third strand down, plaited again, with the third strand up, and then it was tied, rolled up and tied again with string. By 8am, both ploughman and horses would have had breakfast, and the horses were taken to the water trough, or, most often, to the bend in the road where there was easy access to the burn, and then straight to the field to plough. By about 10 o'clock, the seagulls and crows would arrive to feed on the upturned sods, and within an hour, there would be hundreds skirling and fighting for position nearest the plough to get the best pickings as the mould-board turned over the furrow. Often you could feel their wings hitting your back or legs, as they dived in behind you. After the feeding frenzy was over, they would settle in a group farther up the field, and rest for an hour or so. As the light faded, they would take off in two's or three's, flying low over the ground for a few hundred yards, then rising steeply and disappearing to their night-perches.

A long day behind the plough would be an eight hour shift, with an hour for lunch. In that case, both ploughman and horses would come back to the steading for lunch, and the horses would be unyoked. On a short day's ploughing, lunch was in the field, and the horses could either be fed in the field, or might not be foddered until night. As children, we used to take tea out to my father during his breaks, in tapered enamel cans, with metal handles, whose lids served as cups.

An opening would be made in the field by using poles, or, quite often, large stones, which would be placed in a line, in order to allow the ploughman to plough the first furrow straight. When standing in between the stilts, he had to be able to look up the field, and keep the poles dead in line. At this point, the horses had to be guided by the mouth or by rein, since they had no furrow to follow. Once the first furrow had been made, one horse would walk on the land and one in the *fur'*, or furrow. At the top and bottom of every length, the area left unploughed all the way round the field was the *head-rig* or *end-rig*, which was left unploughed until last. At Torhousemuir, of course, there would be only one pair at work, but elsewhere, when I began to work out on other farms, two teams could be ploughing at once in the same field.

When you made the opening, you would keep turning to the right, on a ploughed area which got wider and wider around your first furrows. This was known as *gathering* and you would gather for a distance of 10 yards. The two initial furrows lay against each other in opposite directions, and were called the *crown furrow*. From the centre of the gathering, you would then step thirty yards to begin another opening, leaving the space in between unploughed. You would repeat the process of gathering for another 10 yards, and then begin work on the area in between, a space of twenty yards. In this area, you would *scale*: that is to say, plough the intermediate twenty yards by turning left and thus turn the sod outwards. The last furrow, which

was an open one, with the two furrow slices facing away from each other, was always the most difficult to throw over. It was called the *hinten* furrow. The two movements of gathering and scaling created a rig of thirty yards; that is to say, thirty yards between each opening and each finish. The field was only roughly paced, and yet experienced ploughmen could work accurately to within inches of the measurement of the field. In lea fields, the head-rig was, I think, ploughed working from the inside out, turning the furrow towards the inside of the field, whereas on stubble fields, the ploughing was from the outside in. This had to be done on alternate years, throwing in one year and outwards the next, or else a hollow would be made next to the fence, with the repeated ploughings.

Though the ploughs may look like simple machines to modern eyes, they were capable of a great deal of flexibility. The adjustments to the width of cut were made on a series of holes on two plates: vertical adjustments determined the depth of the cut, and the horizontal adjustment determined the angle of the plough and the width of the cut.

When ploughing the lea, the plough was adjusted to open about seven inches of earth; in stubble the furrow was wider, as I said: as much as nine or ten inches wide and six inches deep. When ploughing the lea, the width

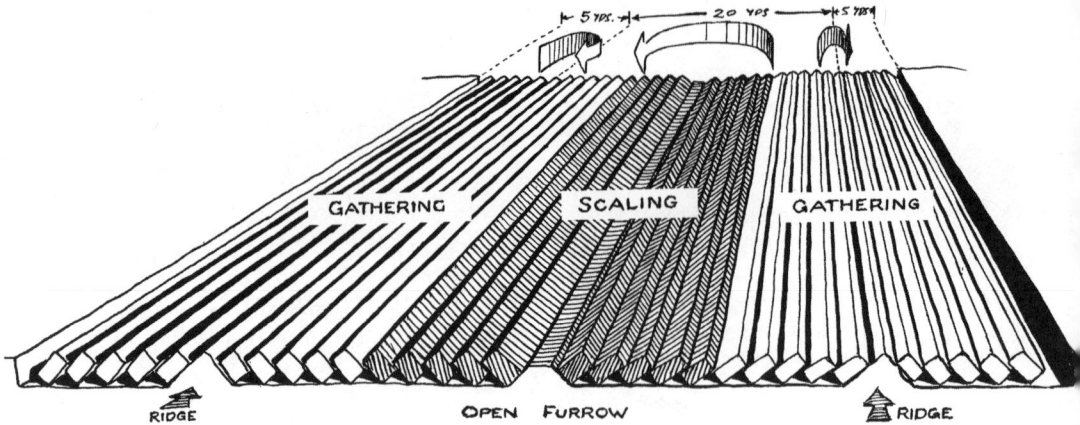

of the cut determined how far apart the seed-beds were: because of the way the furrows lay against each other, the seedbed in the hollow between the ridges would be about every four inches. The small furrows dried out more quickly and enabled you to get the seed sown earlier. When the seed was broadcast, it would run down into the little ridges created by the plough, and the crops would grow in straight lines.

Horses pick up on your moods, and the movement of their ears shows that they listen to whistling or singing. Once upset, however, they take time to settle back down. There were two commands when ploughing: *hup* was the command to turn right, and *yean* or *whine* to turn left. *Hup in* repeated two or three times indicated that they should turn sharp right at the end of the furrow. At the end of the furrow, you had to lean down on the stilts, so as to keep up the beam, or head of the plough, so that the horses did not step over the slackened chains as they turned sharply to begin the new furrow. If they did so, they were sensitive enough to know that they had done wrong, and this would upset their rhythm, especially if the ploughman raised his voice in anger. The horses repaid kind and quiet treatment, and my father often assigned them to me, since I liked the work and respected their intelligence.

As I learnt, when I later went to work on a neighbouring farm, young horses had to be broken to the plough. They would be tethered to an older horse, and then yoked up to a heavy log, being driven round the field until the horse became accustomed to pulling a heavy weight and being driven on long reins. When the young horse was introduced to the plough, it was yoked along with an older horse, until it settled in to being driven and obeying the spoken commands. When the horses were worked daily like this, they soon settled to a steady pace, and could work from 8am until 5pm without strain. When ploughing, too, the younger horse would be yoked to a chain shorter by about two links than the other horse's chain, so that it would pull more and tire it out more quickly. It was a way of introducing them to a pace they could keep up all day, and a ploughing team proceeding at a steady pace would achieve more than one which set out too fast and got tired out more quickly. When the horses were brought home at night, one of us would ride on the back of one, sitting side on.

Ploughing began in the autumn, and carried on until spring. In the wet weather, the heavy sticky soil, which we called *claggy*, would cling to your boots. The horses could not work in really wet weather, since they could not tolerate rain in their faces and would try to turn. It was only at night you realised how wearying it had been. In spring, the cold dry winds would cut through your rough trousers and chafe your thighs: as we would say, the skin was *piskit*. Your hands would crack open at the joints, mainly on the inside, and we would rub them thoroughly with Vaseline to try to keep the skin soft. In really hard frost, you had to abandon the effort altogether, until the thaw.

There were frequent ploughing matches round Wigtown, which we attended, and which were some of the few occasions on which crofters like my father took some leisure time. It was only after the war that the first tractor was entered for one of these events. There were a variety of prizes to be won – for the oldest or youngest ploughman, or even for the one *wi' the maist weans* – which made them popular events.

SOWING

So the grass field, or lea field, had been ploughed during the winter months. When it had dried out in spring, it would be prepared for sowing oats. Elsewhere, disc-machines or corn-dibblers, drawn by horses were already in use, but at Torhousemuir, we were still using the broadcast method. In preparation for sowing, bags of seed-corn were placed at intervals in the field. Someone would be responsible for carrying seed from the sack to the sower. The oats were sown from a *broadcast sheet*. This consisted of a kidney-shaped frame, designed to fit round the sower's stomach. It might be a wooden frame, about 2 inches by half an inch, with sacking or hessian sewn onto it. In fact, we made our own out of old bicycle wheels, bent inwards to create the right shape, and the spoke holes were used for anchoring the sacking.

The sower had a rhythm, marching up and down, following the straight furrows, and throwing the seeds in a wide, slow arc, to ensure that they landed with as even a spread as possible. While one hand cast the seed upwards in an arc, the other hand would dip into the sheet for another handful. Smaller steps would ensure a thicker sowing, while long strides would sow the crop more thinly. The same adjustment had to be made for going up or down hill: larger steps had to be taken up hill, and smaller ones down hill to keep the dosage even. Men were so practised in this method that they could step a field precisely, to an exact measurement of seed. This method was used for corn, also for fertiliser, and, when we first moved to Torhousemuir, for grass seed. On neighbouring farms, manure was being sown by horse-drawn machines. The difference in sowing grass from sowing corn was that one handful would last for three throws, so that it required some skill to dispense only the correct amount each time. Later on, for grass-sowing, we acquired a seed-fiddle, which consisted of a hopper made of sacking and fitted to your shoulder, which ran down to a circular sprayer. Below that, and attached to its spindle, was a bow of wood, with a strong cord, which could be pulled left or right, turning the sprayer this way and that as you walked.

If the crop was oats, which was to be followed by a grass crop, the corn was sowed first, and a fortnight or a week later, before the corn sprouted, the grass was sown. This was given one run of the harrow and then rolled.

Harvesting.

HARROWING

BECAUSE of the introduction of the mechanical reaper, the ground had to be prepared to ensure that nothing would harm the machine. The furrows were then harrowed to cover the seed, and the stones were removed by hand and carted off.

Harrowing was carried out several times across the field at different angles. The harrow consisted of a criss-cross of about eight feet wide in a diamond shape, with teeth every four inches; there were also zig-zag harrows, but we used the diamond harrow. It dug in to the top of the furrow just enough to displace the soil and cover the seed. The sweep of the harrow naturally covered a very wide area, as compared with the plough, which cut only seven inches at a time, hence the old expression: "When you can harrow what I have ploughed..."

The roller we had was brought from Killiemore, an old crofting settlement by Kirkcowan. It was made of granite, and we brought it along the Kirkcowan road roped to the back of a cart. It was about 5 foot long by 14 inches in diameter. Later on, the rollers were iron. Rolling was a one-horse operation, and the purpose was once again to ensure that the field was flat for harvesting by a mechanical reaper. You could sit on the roller while it was being pulled, by placing a board to one side and leaning against it.

OAT HARVEST

OATS were generally ready by the middle of August. Quite often a few early cuts were taken with the scythe by local farmers before Wigtown Show right at the beginning of August, so that the Stranraer farmers attending the Show could see how well on their Machars colleagues were with the harvest. It had to be admitted, however, that the oats were generally much too green to be cut.

The sort we grew generally were known as *potato oats*: these did not have a strong straw, but produced oats in plenty. Another variety we grew was *Onward*, which was a good producer, with reasonably good straw. There were varieties, such as *Star*, which produced long straw, but which also tended to collapse in heavy rain.

To *open* a field, you would take a swathe with a scythe, right the way round the perimeter, or head-rig, and also where you entered the field at the gate. This was known as *opening the roads*. This was sufficient to allow free access for a horse without trampling down the corn and it was simply designed to avoid waste. On bigger farms, the delay entailed by scything was not welcomed, so the practice vanished. When the combines finally came in, the cut was to the front, so that the problem was less critical than for the reaper, pulled by horses and with the cutter to the side. The other occasions on which a scythe was used was when the ground was boggy, or if heavy rain had flattened the crop. In all cases, it was done so as to avoid losing any corn or straw.

It was a two-horse reaper we used; our acreage could not justify us in the cost of a binder which saved on labour, but required three horses to pull it. The reaper required two men to operate it: one to drive the horse, and another to operate the tilting rake and the tilting board, shedding off the sheaves in the size required. The horseman was seated in the centre of the machine and the tilter sat over the right-hand wheel.

The reaper had a four foot cutting bar, and the cutting blades within it were driven by the land wheel; the bar was lifted or lowered by the operator, using a lever. The knives going through the fingers made a distinctive rhythmic clicking noise. It was a fairly simple mechanism and could not operate if it was raining or wet; another hazard was that the driving rod was made of wood and could break. A fairly smart pace was required for a good cutting action, and this also kept the horses from turning aside and taking a bite of corn; failing to keep control of the horses' appetites was my first mistake as a horseman. The blades had to be sharpened every night, as part of the ritual of harvesting, so that you always had a sharp set, if anything went wrong, such as a broken blade: I remember father clamping them to a gate to sharpen them with a file. The tilter used his foot for controlling the tilting board, selecting the size of the sheaf by eye. Perhaps three or four cuts would be enough for a sheaf, and he held the tilting board all this time, until finally he released it. The tilting rake had a head at a 45 degree angle, in line with the tilting board.

When I was about age 12 or 13, I followed in my brother Willie's footsteps and I began driving the horse, while my father controlled the tilting board. I had been leading horses probably from the age of six, and most children on crofts or farms learnt early on to drive carts. My father and I would cut one full length of the field; then we would all be involved in lifting and tying the sheaves and putting them to one side. Straps were made with a handful of corn, often by the children, and were used to tie the sheaves, and then adults would lift, tidy and then stack them to one side. My mother would have begun tying at one end of the row, and my father would start at the other end, once he had finished cutting. For a break, we might take a drink to the field consisting of a gallon can of water, with a handful of oatmeal in it. After one length had been cut, another cut would be begun and the whole process go on until 4pm when my mother would leave the field to begin preparing the supper and the milking. Father and I would continue until 5pm.

After supper, we would go back and *stook* what we had cut that day. The stooks were formed according to a deliberate plan: six rows of sheaves were taken in together to form a stook in the middle. On occasion, if the season was particularly wet, we would have to form *rickles*, which is to say, gather four stooks (about twenty four sheaves) into a pyramid to try to dry them off. The sheaves were kept as upright as possible, and there was an open space in the middle to let the wind through. The top sheaf was put in upside down, and its end spread out and smoothed downwards, like a thatched crown, with a band tied around it. The sheaves had to be spread out again before being taken to a stackyard. Fortunately, there were only about two seasons at Torhousemuir when we were forced to do this, because it involved a good deal of extra labour.

Generally, however, about two weeks later, the stooks were usually aired and well-dried and they could be gathered in directly without further stacking. If they had stood for a while, they would be tipped towards the sun to dry out. There was a method to harvesting which was calculated to save on time and labour: for example, the cart would be led up the middle between the rows of stooks, so that the cart could be loaded without walking long distances. Children were responsible for leading the cart from one

stook to the next. The sheaves were forked then onto the cart by my mother. I would generally be charged with building the cart and taking it to the stackyard. The cart would be loaded with five rows of sheaves, which counted as a full load, and it was then tied down before being taken to the stackyard: two hanks of rope were tied to the axles, looped through the corn-rails and thrown over, before being pulled tight on the shaft of the cart. The whole process of loading a cart took about twenty minutes.

As I said, our stackyard was at a distance from the croft, to enable the steam thresher to reach it, since it could not negotiate the bad turn at the bridge. As I unloaded and forked my cart load to the stack-builder, someone with another cart would go out to the field, so that the sheaves were being loaded in a continuous stream. The yard would probably have been empty by harvest-time, since all the straw stacks from the preceding year would have been used up for fodder, though at some farms you might get a surplus from the preceding year.

The process of bringing the sheaves into the stackyard was known as the *lead in* and bringing in the harvest was one of the great events in the crofting calendar. The days of the harvest *kirn* were probably dying out by the time I was growing up. In any case, these would have been more common on farms where there were plenty of workers than on the crofts where mostly only family members worked. When I was older, I helped out with the harvest at Torhouse farm and I remember some sort of harvest celebration: the last man with the last sheaf brought into the stackyard had a bucket of water thrown over him. The kirn consisted of a meal for the workers, followed by singing and a dance, which was held in the barn at Torhouse.

BUILDING A STACK

THE stacks were circular and built from the centre outwards, on top of a *stack-but*. The stack-buts were formed with stones, then with branches and bracken, to stop the bottom of the stack from drawing damp from the ground. If there was a good stack-but there was little waste. When building, the main principle was to ensure that the middle of the stack was high or *well-hearted*, so that water could not enter the stack. You began with a stook in the centre, and the sheaves were stacked at an angle around this, with the corn facing inwards and the sheaf-but facing outwards. The man or boy forking or throwing to the stack builder always ensured that the sheaf landed bottom outwards, so that no time was wasted turning the sheaf. The straw had always to be the part to touch the ground, since any corn affected by

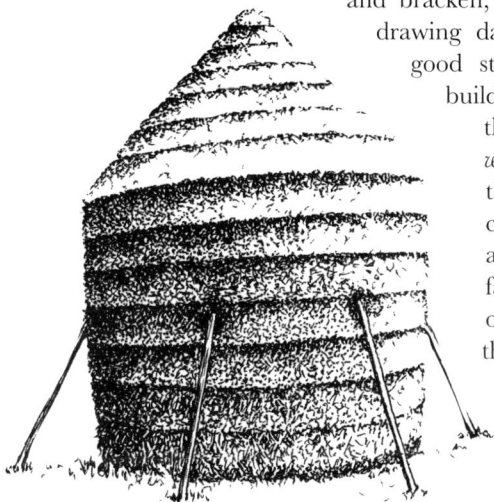

damp or water would grow or rot. Each layer of sheaves, as you worked your way round the centre in concentric circles to the outermost layer, would lie half covering the previous inner row, probably roughly to the level of the strap half way up the sheaf. In this way, the corn always lay on the straw of the previous sheaf, and this layering was continued until you reached the edge of the but. It was work carried out kneeling, and your trousers wore out doing it; I can remember spending the rest of the evening picking thistles out of my hands and knees.

Once you had finished the bottom layer, you began at the outside and worked your way inwards. In the formation of the stack, there were probably about five circular rows from the centre to the edge. The layer on top of the first would be built inwards one foot from the edge, resting on the strap of the preceding layer, so that the building proceeded in reducing circles. The process was repeated until you were satisfied that you had reached the height required for the stack. Towards the top of the stack, you would let it grow outwards a bit, and then took it in for the head: slowly at first for the first three rows, then sharply for the last five rows. This would give a good curved head on the stack: with the angle of the first three rows, it meant that rain dripped off clear. A long head on a stack would mean that the wind would blow against the thatch and flatten it, rather than lifting it up. The crown would be finished with a sheaf laid in upside down, with the but spread out all round the crown.

On occasion, when the season was bad, and there was likely to be dampness in the sheaves, we inserted a sack of straw at the centre, and pulled it up as we built to create a tunnel or hollow in the middle of the stack to aerate it. This had the potential, however, to create a downward slope in the middle of the stack, so risking water penetration, and I do not remember that it was often attempted.

My mother's job was often to fork from the cart, and my father would build the stack – that is, until we were old enough to help with the building. Then, there was a fair degree of competition between us as youths to build the best stack. The appearance of the stack was considered of prime importance: we knew farmers who would rake and trim the sides with sheep-shears. Farmers were judged on the quality of the men they kept, and a neat stack would be a sign that the farm employed a capable man. On the whole, isolated as we were at Mossend, the most we did was to trim the thatch round, so that it had a neat edge.

Stacks could be thatched in several ways: occasionally, we thatched with bunches of straw, which had to be *drawn* between the hands, so that the short straws fell out, and until you had pulled all the stems straight.

Alternatively, reeds could be got from the Mains of Penninghame, where there was boggy ground. Usually, however, the stack was thatched with rushes, from the nearby marshy ground on our own croft: these were cut with a scythe, bundled like sheaves, and carted to the stackyard. We would put a ladder up against the stack and anchor a fork up on top, ahead of where we were working, to support a supply of rushes. Then you would lift up each sheaf, and poke in the top end of a handful of rushes into the sheaf, leaving the wider end hanging out. On the sheaf above, the next lot of rushes had to have about six or eight inches cover over the first lot, in a technique much like that of slating a roof. It was a job which could badly scratch your hands, especially where sheaves contained a lot of thistles, and you might spend the evening squeezing thistle points out of your hands. You would work your way up a segment or *bay* of the stack head, until you had thatched all you could reach from that position, and then change the position of the ladder. It could take one man three or four hours to thatch a stack.. A crisscross of sheaves of rushes or straw tied together with string would be set neatly on the top. The rushes were straightened all down the stack-head, using a stick, and the eaves of the stack trimmed round.

Then the stack was roped down, using six or eight grass ropes which went right over the top; this part of the work was also carried out from a ladder. The ropes were called *bay-ropes* and there would be about four of them, creating eight bays; elsewhere, I believe they were called *ower-gauns*. They were neatly criss-crossed on the top of the stack. Binder twine or possibly grass rope was then woven over and underneath the ropes to form a net: the circular movement of weaving, which went from over to under, was designed to economise on effort - it was the quickest way of moving from one vertical rope to the other. There would be about 8 inches between the lines of twine. All these ropes had been prepared in the winter: *rowing* ropes round your arm was a traditional indoor wet day job, in preparation for the days spent thatching. The ends of the bay-ropes were then weighted with a brick or stone to hold them down. The thatching was surprisingly effective in keeping the stack very dry until threshing day. Of course, if the stack was not properly dried out, it became heated, discoloured, the corn tasted bad and even the cows might refuse to eat it. If a stack was *het*, that was a sure sign of bad management and that was certain to be noted. If stacks were left for more than a year, you might get mice and rats penetrating them, or birds burrowing into the thatch because it was becoming brittle. This rarely happened on our croft, since we had few surpluses.

The stacks were propped with long branches or poles to allow them to settle straight; the props were moved according to which side leaned. By the end of this whole process, we would have about five stacks in the yard, about 8-10 feet in diameter. They were positioned so that the threshing mill could be driven in between the stacks; the idea was that you could fork four stacks onto the mill without moving the machine, because it was a cumbersome operation to set the threshing-mill in the right position, and line it up with the steam-engine.

THRESHING

The end of harvest would be in September; in November, we would expect a visit from the threshing-mill. The one we used was owned by a man from Killantrae Croft, Port William, called Tom Crawley. The threshing mill was drawn by a steam-engine, which pulled it by means of a draw-bar; behind the mill would come the *living van*, in which the owner and the second man would have their quarters while travelling between farms. The living van, which was a double timbered, four wheeled van, would be left at our road end. It had a curved roof, two small windows and three or four steps up to it. Inside it had two bunks, and a coke stove. Here they could wash and change in the van for going out at night and weekends. The men often carried bicycles in the van, which would enable them to go home at weekends, but during the week they lived together as travelling companions and as work-mates. The mills probably worked for about nine months of the year, and since the men were fed away from home, this saved the family's expenditure on food. They would finish by Saturday lunchtime to return home, only to begin again on Sunday evening.

The whole procession – steam engine, mill and van – travelled at about fifteen or twenty miles per hour, and created a great noise as it went, since the vehicles were all iron-shod at this time. For some reason, the mills, which were wooden built, were always painted red, or pink with red uprights. The manufacturer's name was painted on the shelvings which were put round the top of the mill, while the name of the engine owner was written above his cab.

We would contact the mill-men in advance to ask them to come, and the three crofts at Torhousemuir which had their threshing done by Crawley's mill would have it completed in about one day each: each might have had about four stacks to thresh. These days were known as *mill days*. When the mill came to the estate, this was the only time of the year when we had coal about the place; as I said, it was never used for our own heating or cooking. We would order coal and pick up about five hundred-weight of it at Wigtown. You had to provide fuel not only for the threshing, but for the engine to move on to the next farm, so that the expenses of transporting the mill were always paid by the farms. We also provided the binder twine, which was used in the buncher.

Two men came with the mill – the engine driver and the *second man*. It took far more manpower than this, however, to complete the process, probably requiring about ten men to do the work. Neighbours turned out free of charge to help fork onto the mill, to build the straw stacks after the corn had been through the mill, and bag and store the corn in the barns; probably the only paid labour, apart from the mill-men, were the *lousers*, those who cut the bands on the sheaves with a knife, and handed them to the mill men. Usually, the lousers were women, and there was one horrific incident, which was much talked about, when one woman had been frightened by a mouse running out of a sheaf, slipped and got caught in the mill mechanism. There were, of course, plenty of mice in the stacks, and at the bottom, there might be rats. By the war-time, it became compulsory to place netting round the base of the stacks, so that rats could be trapped and destroyed.

There was a whole routine to establishing the mill in position. It had to be towed into position between the stacks and jacked up level. It usually was placed in position at night, when it arrived from another farm, and this might even be in the dark. Travelling at night must have required a certain skill and good eyesight, since it carried only two lights hung on the front of the engine, with a red light dangling at the back. We regularly repaired all the potholes in the road for its coming, not so much to help the driver, as because the weight of the engine would compact and crush the stones well into position and make necessary repairs to the road. Once in position, the mill had to be levelled so that the riddles operated properly. There were shelves on three sides along the top which opened out to form a wider platform for working, and these were always folded in on completion and the whole top sheeted so that the riddles did not seize up with moisture. It was a regular occurrence to have to stop threshing and sheet the whole thing if a shower came on, and work would not begin on a wet day.

The whole mechanism was belt-driven from the fly-wheel on the engine, connected to the drive-pulley on the mill. Smaller belts were connected to the fanner, the riddles and for the buncher. The threshing mill hummed and the engine chugged; even at a distance, you would hear the whine of the mill. The clouds of dust coming off the corn and chaff, particularly in the days before the chaff was funnelled clear of the machine, made you glad of the tea-breaks.

The mill day started at 8am, and the steam whistle was sounded to indicate that, if you were not already there, you had better hurry to the stackyard. On a croft, the mill day might only last until 3pm, but at farms, it might go on until 5pm. At noon, the special dinner would be laid on at the house. For this, my mother would have help from a neighbour or relative. There was soup, potatoes and beef, followed by custard, rice and tea. Local gossip would be exchanged at the table and probably a few tall stories; sometimes the mill men would return to a farm at night for dinner and for breakfast in the mornings.

On threshing day, the reed thatch, which came off the top of the stacks, was first set aside for animal bedding. The corn sheaves were fed onto a drum, which had rotating spikes to knock the corn off the straw. It then passed over riddles, the good corn being separated from the seconds, and the seconds from the *gill*, the weed-seeds and rubbish. The corn was discharged at the end of the mill, and the gill onto the ground at the side. There were two spouts for the corn, and one would be closed off as a bag filled from the other. As soon as that was filled, the other had to be opened, since you could not let corn build up within the thresher. The chaff was blown free by fanners and discharged at the side. It was usually children's work to keep the chaff clear of the underside of the mill. The straw went to the opposite end of the mill, being pushed towards the buncher by kickers, and bunched when it got to a certain weight. There were also the *shorts*, or small broken straws, which came out beneath the buncher on a shaker.

Two people would fork from the stacks, standing on top of them. The engine-man and his mate fed the mill, pouring into the mechanism the sheaves, which they caught from the *lousers* or those who cut the sheaves. The lousers handed the sheaves down to the feeder, who was about a foot lower than them. Lousing was hard work, since not only could the sheaves be almost half thistles, but because, until the sheaves piled up to waist height, the lousers were permanently bending to pick them up. Another person would build the new straw stack, and one would fork to him, so that the process of dismantling a stack while a straw stack was built up nearby was one continuous motion. Two men would meanwhile be bagging corn and carting it home. Children would run to fill the tank on the engine, located just below the driver's compartment. It had to be filled constantly: I can remember peering over, thinking it was full, and then the driver would suck it out, and the whole process would have to begin over again. In between, the chaff had to be raked from beneath the mill, and it would be taken away on a sheet, made of an old bag split open. You had to put up

Threshing.

with the mockery of the older men: "I could carry more in my boots", they'd say, as you ran with all your might. Later on, that job was accomplished by a funnel from which the chaff was blown clear of the mill.

The straw which had been in the buncher was now in much larger bunches, which were no longer called sheaves. The new stack was a *sow stack*, so called because it was long with rounded edges. Four or five stacks of corn would now be built into one stack of straw. The ends of the straw were raked downwards, so that they acted as thatch. When using the stack, you would take straw off the end, so leaving the rest intact, because it had to last all winter; a cart would take about sixty bunches. On occasion, we might have to buy straw in from the neighbours, particularly if we had extra cattle. We probably brought in straw from the stack every two days, and kept some stored in the byre as fodder for the horse and cows.

The corn had to be put indoors before night. In our case, it was loaded by the cart-ful, in ten-stone hessian bags which had been saved for the occasion. In the barn, next door to the house, it was stacked about three sacks high, in rows separated by a space, in order to keep the mice down. Boards had been laid down on the concrete floor to prevent the sacks from drawing damp. The barn was also used to store the straw which was in immediate use during the winter. The walls of the barn were finished at eaves' height, so that it was an easy job to throw the straw over into the byre.

The corn had been separated by the threshing process into firsts and seconds, or *tails*. Some of the firsts became the seed-corn of the next year, and the seconds were used for feed for hens. The corn used for animal fodder was usually bruised, at a neighbouring farm. The bruiser consisted of a hopper and two iron rollers, which crushed the corn as it went through and discharged it onto the floor. It was powered by a petrol engine, and we probably only paid our neighbour for the fuel. We did eventually acquire a hand-bruiser, and finally a petrol-powered machine. Good clean chaff was used for mash, which included chopped straw, bran and watered-down treacle, and which was fed to the cattle. Sometimes, chaff was kept in a bag, to be used for bedding, if it was clean and clear of thistles; potato oats gave the best chaff for this. *Cauf-beds* were very warm, because you sank into the hollow. With us, these were used only as temporary bedding, which was put out on the floor.

Hay Harvest

Hay was reaped in June or July, using the two-horse reaper. No tilting board was used because there were no sheaves to be gathered, so that only one man was required to operate the reaper, and it was usually an easier process than the oat harvest because the weather was generally fine. There was a divider board which pushed the cut grass away from the standing crop, so that the knives did not get clogged. The machine left it lying in neat long rows.

Hay harvest was usually a time when you saw the partridges, which often nested in the hay-fields; they were very common then. The driver would probably spot the nest when driving past it on a previous cut, or have noted the alarm of the birds above their nest. It was quite usual to stop the reaper, put the nest to one side, leave a small patch of hay for cover, and replace the nest behind the machine. Usually at this time, the eggs were nearly ready to hatch.

When the swathes of cut hay were dried on top after a couple of days, they were turned by hand with a wooden hand-rake which had wooden teeth or else a two-pronged hay fork, to allow the underside to dry. Since the hay was laid down by the reaper in a certain direction, the turning operation was a simple matter of reversing the direction by hand. At the top of the field, you would change direction, and you were then working against the run of the swathe, so that you had to adopt a different stance. The hay would be pale on top and bright green underneath when first turned. Then the two rows were raked into one by hand; some farms, of

Building the hay rucks.

course, had horse-drawn hay rakes and swathers, but on the croft it was all carried out by hand. Sometimes, on the croft, if other work were pressing, only a certain amount of hay was cut at a time, the rest of the grass being left until a more convenient time.

When the drying process was complete, which usually took about two days, the hay was swept into heaps by a horse pulling a sweep. The horse-drawn sweep was about eight foot wide, and had six tines which collected the hay. You had to control the angle of the tine with care, so that it did not skip over the hay. Once you had enough hay gathered, a handle was lifted, the tines caught in the earth, and the sweep would tip over. About three sweepfuls were used to build the *rucks*. If, however, the weather was poor, the hay was put into small heaps, *coles* or *coils* which gave some protection to the hay and enabled it to dry more quickly, for putting into rucks. The coles consisted of about two good forkfuls of hay, which were dressed down so that they would shed the water, and given a rough raking down with the fork.

A ruck was about 5 foot wide by 7 foot high. You had to clear the slack hay from the bottom of the ruck by hand, or else the loose hay would be wasted. The whole field would also be raked by hand, to gather together anything raked off the rucks, and the waste was raked about four feet away from the rucks, to enable the hay-sweep to gather the last of the surplus for building into the rucks. The hand-raking would probably be done in the evenings. Someone, usually a child, helped tramp the rucks down as they were being built. When forking hay onto the ruck, you had to turn the fork so that its tines were downwards and it could be easily pulled out of the hay. It was, on odd occasions, a hazardous business for the ruck builder, and I remember getting a tine through my foot. The ruck would stand for some time – perhaps up to four weeks - and it might be carted home either before or after harvest. In the interim, we would sometimes rope the rucks down: a handful of hay taken from the base of the stack, was twisted, tied to a rope, and then, when this was flung over the top, it was turned round another twist of hay on the base at the other side. Two ropes were used on a ruck and would be enough to keep it from lifting in the wind until it settled down. Ideally, the hay would be brought in before the harvest, but this would depend on the conditions, and it was one of the sights of the countryside during July and August to see the fields full of hay rucks.

When the rucks were ready for stacking in the yard, hay was forked from the ruck onto the cart by hand since at the croft, we had no rick-lifter. There was a system to building the cart, beginning with about three or four rows round about the outside of the cart, and then filling up the centre. The builder standing on the cart would receive a forkful, and roll it so that the straggled ends were kept within the cart, carefully placing it in the right position. Two rucks would fill a cart. Once loaded high, we would sit on the top of the load and take it home to the stackyard.

The haystacks were built in the stackyard at home, in front of the croft, since there was no milling involved and no need to consider how to get the mill to the stackyard. As with the corn-stacks, the hay-stacks were built on a but of branches and stones, which you filled out. The process went from the outside of the but, and then to the middle, which had to be kept high. The stacks were not built as high as corn stacks, and only a small area of the crown had to be thatched with rushes: the hay could be raked easily to form straight lines, which would shed the water.

It is hard to over-state the importance of bringing in the hay and harvest: upon its successful completion, the whole economy of the croft depended. Poorly dried hay would overheat and might mean that hay would have to be bought in, using up valuable cash reserves. The stacks were eventually opened and were used as fodder, mainly for the cows; the horses normally were fed straw. When a stack was opened, the thatch was put to one side and used for bedding animals.

ROOT-CROPS

THE root crops were planted on the stubble fields of the previous year. This red-land was clear of weeds and the soil was relatively loose for ploughing and, as I said, we used a plough with a larger cut than we did for the lea fields.

To prepare the stubble field, after harvest, the dung-midden was emptied and the dung spread on the field. The dung was barrowed to the midden daily, or twice daily when, in the winter, the young calves were lying in a great deal. Sometimes the dung would be spread direct from the cart, or else it could be dumped in heaps, which would form neat rows, with spaces between the heaps of about 5 yards, and then *scaled* at a later date. In this case, you would lead the horse five paces and draw it off the cart with a long-handled curved fork or *graip*; it was also known as a *dreg*. If only a light covering was required, you might take six or seven paces before drawing off the next heap. The cart probably held about five or six heaps at a time. For this job, you would tie some sort of protection, quite often just strips of hessian sacks, round the bottom of your trousers and tops of your boots. The dung was ploughed in during late autumn or early winter, during what was known as *the back en'*, before the ground was worked for the root crop. Dung would also be used on the young grass, which was to be grazed after harvest and then grown on for a year for hay.

For root-crops, the land had to be grubbed in spring to give enough tilth for the seed. The grubber was about four to five feet wide, with three wheels. There was a frame with bars across, and six tines fixed to this dug into the ground, breaking up the furrows. There was a lever to regulate the depth of the digging. It was a two-horse job and it might even be carried out twice. After grubbing, the plough furrows vanished, and a drill plough was used to form drills for turnips and potatoes. Immediately before drilling, manure or inorganic fertiliser was sown by hand; for potatoes, it was sown over the drills.

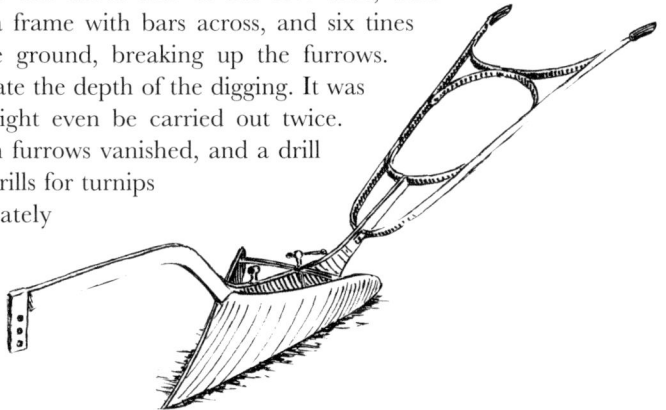

The drill plough consisted of two straight mould boards, as distinct from the ordinary plough, which has one curved mould board. There were no wheels on the drill plough, and it simply had to be balanced by the ploughman. The drill plough simply pushes the soil to either side, and when ploughing down in the opposite direction, the soil would be pushed against the first ridge. The plough had a scratch-marker on it, which would mark out the position of the next drill while you ploughed the first drill: it consisted of a piece of metal, which trailed on the landward side, and, when you turned at the top of the field, could be flipped over to the other side using a swivel mechanism on the top of the plough. The scratch marker was adjustable, and had a pin at every inch to allow different lengths. The drills were always calculated to suit the cart, which would be used during the future harvesting of the crops and whose wheels were forty-eight inches apart.

POTATOES

For a potato crop, if the land had not been dung-ed the previous autumn, dung was spread by fork in the new drills, and the potatoes planted directly into the dung. The cart would be led along the field, its wheels in the hollows which had been opened. The seed potatoes were left in sacks at intervals along the field, and we filled our buckets from them, planting the potatoes from ten inches to one foot apart. The drill plough would then be used again to split open the ridge, to form another ridge covering over the potatoes: that is to say that two horses pulling the drill plough would walk along the drills, splitting it behind them, to cover the potatoes in the *heugh* or hollow. The horses would have to learn this technique of *topping the drill*, or walking in between the lines of potatoes, on top of the ridges of soil. Sometimes, fertiliser would also be spread by hand, using the broadcast sheet.

During the season, a one-horse machine called a drill harrow would be used, to keep the weeds in the *heugh* down, and they would be drilled again to build up the drill, or *hap up* the potatoes.

In October, prior to potato-lifting and when the *shaws* had died down, the drill would be split open with the plough, which would push the potatoes to the side. The horses walked in the furrows between the rows of potatoes, pulling the plough behind them. The new furrows still had to be scratched down by hand, to check for those remaining below the surface. The ground was also harrowed to bring up any which were left covered and stubborn shaws were taken away for burning.

The potatoes were lifted by hand and put into a wire or *spale* basket, or a bucket, and thence tipped into sacks. At other farms, squads of workers would be employed, and the October holidays from school were known as the *tattie holidays*, and were timed to enable children to help with the potato lifting. Generally, we used buckets, because the baskets required more than one person to lift them, and there were none of the large squads of lifters on the croft which would be employed on the large farms. The sacks were only half filled, so that they could be lifted onto the cart. If they were bagged, they were carted home; those which were put straight onto the cart would be taken to a point near the entrance to the field, to a pit or *clamp*. To make a clamp, the potatoes were built up from a base that was about three foot wide and shaped like a long oblong which might be thirty foot long, to a height of a couple of feet. Alternatively, we might have built two, about twenty feet long. They were covered with straw, which was arranged around them upright on the ground, and the top of the mound was neatly covered with another layer of straw. On top of this, we placed earth to a depth of about eight inches; since this was dug from round about, it formed a trench about one foot deep, so that the potatoes stood above the ground level, and water ran into the trench. The clamp was designed to keep them frost-free over the winter. The end of the pit could be opened and re-sealed, according to need. It was a good method of storing potatoes, since sheds could not always be kept frost-free.

The potatoes had to be *waled*, that is to say, go through a process of storing and selection: the small potatoes were kept frost-free for seed next year, the *chats* were simply rubbish, while the *ware* or large potatoes, were kept for eating, or, if there were a large surplus, for the cattle. The chats, however, were not thrown away, but fed to pigs or other animals: in winter, they would be boiled in our boiler with bran or oats, to make hot hen food. The selection process was, in our case, carried out by hand, but on larger farms, there would be riddles, turned by hand, with different sizes of meshes, which would allow the different sizes of potato to drop through.

During the war, there was a great increase in potato growing, because subsidy of £10 per acre was paid: surplus potatoes could be bought back from the government, sprayed with a blue dye, and were intended to be used for cattle feed. Locally, however, they were bought for eating, since the dye rarely went beyond the top layer of potatoes.

TURNIPS

TURNIPS were the most important crop for providing cattle food; silage was unheard of in the area until the 1960's, so that turnip was the key to keeping the cattle in feed over the winter, and was grown in order to help increase the milk yield.

Turnip seed was sown using the turnip seed barrow, which was equipped with specially shaped rollers to go over the tops of the furrows; they were adjustable, so as to fit the width of the drills. The barrow had two seed-boxes, one positioned above each row. Inside each there were revolving scoops and a brush to keep the aperture clear and the seed running out smoothly. The box had a half-inch hole on the side, and on the outside, fitting over it, there was a brass plate with numbered holes, from small to large. The settings could go as far as number nine, but I remember we generally set it at four. The seed ran from the box down into a tube, and was sown about two inches deep on top of the drill. The soil fell in on top of the seed, as the barrow passed, thanks to a sort of miniature plough which was immediately behind the tube end. These were sown in early spring and became ready in about a month. Sometimes, they were sown at intervals, so that not all the plants became ready at once.

During the growing season, the drill harrow, which had tines on it, was used to keep the weeds down. This was a one-horse operation, and had to be handled with care, since there were only two feet between the drills. The size of the drill was also cut down by the harrow. This was probably carried out in May or June, about a month or six weeks after sowing. There was also *back hoeing* by hand to keep the weeds down. There were, of course, no weedkillers and in a wet season, the field could be full of weeds, since the hoeing could not kill them off; if the weeds seeded, they would also grow in among the corn and young grass the following year.

Between the end of May and early July, the turnips had to be thinned. On the croft, it was work which could be done in stages,

because the planting of the turnips was staggered, so as to allow us time to cope with the hay harvest. Thinning or *singling* was either carried out by hand, or using a turnip-hoe. Children usually thinned by hand, since they could not be trusted to be accurate with a hoe: they tied sacking round their knees, and crawled between the rows. The hoes used by the adults were five or six inches wide, and were used to create a space of six inches between the plants. There was a skill to swift thinning: it involved clearing the sides of the drill of weeds, and clearing right to the root of the plant which was being kept, using the corner of the hoe, so that the plant fell on its side. In a day or two, the plant would right itself. Good thinners, to show off their skill, would ensure that all the turnip plants were lying in one direction in the row. The thinners all worked in rhythm, and on larger farms, the field would be opened by the head ploughman, and the others would follow on, moving in a slanting line which kept strictly in step.

In good soil, without many stones, it would take about twenty minutes to thin one hundred yards, though on piece work, it was often done faster. Thinning turnips was extra work which could be got on neighbouring farms, and was paid by the hundred yards. When I began the work, it was paid at 3d. and by the end of the war, the rate had gone up to one shilling. Farm workers and cottars welcomed the extra work, since the whole family could get involved, and they would often turn out in the evenings to do it. Families would move from farm to farm, as they were required; it was not uncommon to see prams parked at the end of the rows. Each person would get a row to thin, although children probably shared a row.

The *shawing* or *snedding* of turnips was also casual work which women, in particular, would often perform. It was done from November onwards, and was cold, wet and back-breaking work: it involved cutting the tops and roots off the turnips. At least on the croft, it could be interspersed with other work and there was some respite from it. On the farms, it was contract work, which would be paid by the hundred yards. Often, with us at least, there would be two days shawing, followed by a day carting the roots from the field. Two rows of turnips would be tackled at once, using a *snedder* or *heuk*. The worker would stoop in between the two rows, and drop the roots to the right, and the shaws to the left, leaving the turnip in the middle. Coming back down the field in the opposite direction, the same would happen, so that eventually four rows would be gathered into one. The roots had occasionally to be knocked to one side to break the root, and the work might well be carried out in hard frost. By the end of the process, a horse and cart could fit in between the rows. The loading of turnips onto the cart by hand was called *clodding* turnips; it could also be done by spearing the roots with a fork. Those gathering the roots would work from four yards behind the cart up to the front of the cart, loading from two rows of turnips at a time. On steep ground, you could only half load the cart, for the sake of the horse, which might hang itself with a heavy load, as happened with poor Peter Lindsay. Often, on the farms at least, one man would have two loaded carts full of turnips, and he would lead one horse and drive the other, keeping both carts dead in line.

The idea was to harvest the turnips before the wettest of the weather, or before frost. The turnips were tipped out of the cart and piled up in clamps, with a covering of straw or rushes, pointed at the top to shed the water and to keep them frost-free. They were less affected by water than potatoes, so that they had no covering of earth. A sod or two might be put on top to hold the covering down against the wind. Generally, the *pit* would be kept near to the byre for winter feed, although some might be used for human consumption. For the cattle, we had a turnip slicer, which, when you raised a handle, forced the turnip down onto a bed of knives. Later, at Torhousekie, we had a turnip-cutter, which was operated by hand, and had a hopper and drum, which chipped the turnips.

The alternative to turnips, as a cattle fodder, was cattle kale. It had to be cut fresh every day, and in wet weather or in frost, it was a job which meant your getting soaked down one side while cutting. We did not grow much kale on the croft, but, insofar as we did, it was generally used as a feed early in the winter, before using the turnips.

DAIRYING

AT Torhousemuir, all the crofts were dairies, and the milk cheque on the 25th day of each month was the most important source of income. Payment was by the gallon, at the rate of a shilling a gallon. At Mossend, we had between five and seven cows and their followers, which were Ayrshire cows, and, in those days, they still had horns. Only bulls were de-horned, and seeing the operation carried out with a hack-saw at a neighbouring farm was one of the more brutal sights I can remember. My mother milked the cows by hand, using a milking stool, and a *luggie*. This had one handle, and a base suited to the angle at which it was held on the floor. My mother would also milk at other farms: in 1936, she walked out to Auchleand, a mile across the moors, night and morning, to milk, and once was lost in the fog. When milking, she always wore a *dust-cap* or kerchief round her hair, because her head would lean against the cow's side while milking.

In the 1930's, the new hygiene regulations were just coming into force: all the utensils were metal by this time, and had to be thoroughly scalded, including the churns from the creamery, which were only roughly rinsed before being returned to us. If the churns were not scalded, the milk would not keep. We paid a rental fee to the creamery for the use of the churns, as well as a fee for haulage. For scalding, we took buckets of boiling water and soda crystals into the dairy each day; it was only later that we acquired a steaming chest. Sieves were washed in cold water at night, and then scalded again the following day. All the utensils were kept in the dairy, which was a corrugated shed at right-angles to the house and byre, about eight foot square. The inside of it was lined with wood and it housed the milk cooler, the churns and other equipment. There was a drain in the middle of the floor, which ran into a soakaway, and a skylight in the roof.

As part of the new regime in dairy hygiene, we had the wooden stalls removed just after we moved to the croft and replaced by concrete stalls. These were of a regulation width, and had a channel or *griepe* to take the waste away from the stall. The walls had to be cement-rendered up to a certain height and floors replaced in concrete.

The procedure after milking was that milk was taken in a carrying pail over to the dairy; there, the milk was poured into a vessel on the top of the cooler, which was about two foot high and eighteen inches wide. The cooling element below it was cooled by water, which ran through the pipes

from a tank outside, connected to it by a hose. The tank was simply filled by buckets, drawn from our pump by hand.

When the tap on the vessel above was opened, the milk fell onto the cooler, and dripped into the second trough. This might be going on while the milking proceeded, because it was a slow process. The last trough channeled the milk into the churn, on top of which was a sieve which held a cotton pad, and had a perforated plate beneath it.. The milk pads were bought from the creamery and new ones had to be used for each milking.

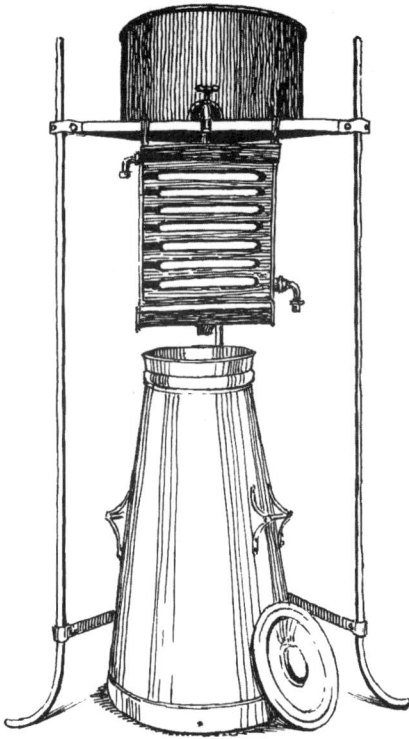

The pads came separated by tissue, which, as I said, my mother put aside for toilet paper. Milk was inspected daily from samples taken at the creamery and sent to the laboratory. A sanitary inspector would occasionally come, if the milk a croft or farm had sent to the creamery had dirt in it. There was still, of course, no means of refrigeration once the milk was in the churns: in really hot weather, the churns might have to be set in a bath of cold water, to await the morning collection. If there was not enough milk to fill the ten or twelve gallon butt, the milk was held over, until it could be supplemented, and would be kept on the croft for two days, but this was usually in winter, when there was less risk of spoiling. The summer yield was about two churns a day, but might be as little as four to five gallons in winter. Later (in the war), there was a premium paid for winter milk, so that meant that crofters would aim to have the cows calve at that time.

We took off some milk each day for our own use and for butter-making. My mother would set milk in a milk-plate at night, and by morning, the cream would have separated and risen to the top. It was then drawn off with the skimmer, a perforated paddle, and used for butter-making, which we did about once a week. The skimmed milk might be used for baking, or some might be returned to the creamery churns, but an excess of skimmed milk would be fed to the calves; if not, the milk sent to the creamery would have been too thin. At first, we made butter in a large wooden churn, but it was difficult to set aside enough cream to make a full batch - at least without complaints from the creamery about the quality of the milk. On the other hand, if you set aside cream over several days, the cream might be sour and affect the butter. Eventually, we acquired a glass churn, which made about one and a half pounds of butter per week. The process was to squeeze the water out of it, and add salt. A butter-pat would be used. For guests, a thistle pattern was impressed into the butter.

The milk was taken to the end of the road each day in the spring cart, where the creamery lorry collected it. We collected the neighbours' milk on the way down the road, and they paid a small sum for having it carted. In the snow of 1937, the roads were entirely blocked, and we could only clear a track wide enough for a wheel-barrow to the main road. The milk had to be wheeled tied onto the barrow, and the lid was also tied on as a security measure, after one of our neighbours had lost his supply entirely when the barrow tipped over. Another method we improvised was to lean the churn on the bicycle pedal, and to push the bicycle at an angle down the road. There was deep snow again in 1947, and this time, was so severe that it threatened the entire livelihood of the crofts. We only had enough churns to last two days, and we were storing milk in all sorts of containers, churning it and setting it to avoid its being totally wasted. It took a combined effort by the crofters at Torhousekie (where we by then were) and Torhousemuir over three days to reach Bladnoch Creamery: we would set off each day with shovels, flasks and sandwiches. We made a track using the fields where the snow was shallowest and finally broke through. When we had made a way passable, I took a horse with a trace horse and some ten stone flour bags, and rode bareback to Wigtown. I filled the bags with two dozen loaves, still warm, and delivered them to the farms and cottages on the way back.

The usual routine was that a lorry picked up the churns at the end of the road. Its route began with the Corsemalzie road – to Kirwaugh, High Barness, Low Malzie, Clugston – and then it would turn back for Wigtown, taking in all the farms on its way. That area was called the *back o' the water*. By the time it arrived, there would be four or five carts waiting, some carrying the milk of two or three crofts. We had to be at the road-end at 8.45am, which meant leaving Torhousemuir by 8 o'clock. The road-end was where the gossip was exchanged, and the post for Torhousemuir was distributed to the waiting milk-carts. The postman was on a bicycle, and this saved him a long ride up the road, especially in bad weather – that is, until one of the farms complained about not getting the mail before Torhousemuir. The carts stood about level with the wagon for easy loading. Occasionally, people would get lifts to Wigtown with the driver in the cab. There were about five wagons collecting milk in the area, all based at Bladnoch, although in summer, when there were surpluses, we might send milk to Dunragit, by a different driver.

On the croft, animals were only kept if they could contribute to the domestic economy through what they could produce. Any cow which was a poor producer would have to be sold before its second year milking. Bull calves were always sold at the market, or to a dealer, and fetched about five shillings. They were put in a sack with the ends tied. Heifers could be kept for stock, since my father always had in mind that we would some day be stocking a larger holding. The cows were brought inside in October, and would go back out again in April. During winter, they were only out during the day for a drink, since the cold could put off milk production and the animals could lose a lot of flesh by being out in the cold.

When the cows were in season, they were taken to a neighbouring farm to the bull. If you waited, you wasted time in getting the cows into calf and therefore into milk. On our croft, the bulls were too expensive to keep, whereas the cows consumed less and could give milk. We paid five shillings for the service of the bull. We would walk the cows down the road, usually taking two at a time, since they were calmer in pairs and easier to handle. Taking the cows away from the bull again was an alarming process and we usually barely managed to get the gate shut in time. We largely looked after animal health ourselves: I remember piercing the animals' side for gas, which came from eating clover, making a measurement from the hip-bone down. As children, we would also curry-comb the cattle, cleaning them of *tartles* or dung clinging to their hair when they had been lying in. We did get one or other of the Wigtown vets in for emergencies, like breech-births, since we could not afford to lose a cow.

SHEEP

WE did not keep sheep at Torhousemuir: the dykes were too poor and not kept up by the estate. I began working with them, however, when I took work on neighbouring farms. By the time I was 15, I was working at Clauchrie farm two or three days a week, whenever there was work, beginning after I had taken the milk to the road-end. I often did not leave for home again until it was 9.30 or 10 o'clock at night. I had to learn on the job, since we had no sheep at home. Generally, my jobs were dipping sheep and lambs, catching for the clippers and helping with the lambing.

In those days, lambing would occur in late March, when the grass had begun to grow. The lambing tended to happen in the morning and evening, and for some reason, there always seemed to me to be more lambs on a wet morning. My first task was to check for any ewes which were having difficulty. The ewes with twin lambs would be taken to the lower field, while those with single lambs would be left on the hill. This was because those

with twin lambs would have to produce more milk and so needed better grass.

In April, there was also the unpleasant task of docking and cutting (that is, castrating) the lambs. Usually two people, the farmer and the shepherd, did the cutting. They would often pull the minute testicles out using their teeth and the blood would run down their chins. All around, the mothers were calling and the lambs bleating. On those evenings, on many farms, a pan of sizzling sweetbreads would be served for supper.

I learnt that there were certain basic signs to follow in the herding of sheep. Sheep which had maggots would be recognisable from the stamping and tail-wagging. In deep snow, such as we had in 1937, 1940 and 1947, sheep could disappear entirely. We knew that they would lie three to four yards from the dyke, keeping out of the wind, so that that was the first place to start looking under the snow. We would plunge a stick down into the snow and twist it, to see if there was wool on it when we pulled it back out.

Late in the season, the sheep would be dipped in a trough seven or eight feet long. Dipping was compulsory, and prevented maggots, *tades*, scab and other diseases. We would plunge them under the solution using a byre brush, and they would then stand in a pen which drained into the dipper. After dipping, the ewes had to be separated from the lambs for weaning or *spaning* and be taken into different fields. We would use dogs to separate the ewes and drive them into a different field. On the day this happened, it was impossible to sleep for the noise from the lambs.

In May, we gathered in the sheep in to the stone built *buchts* for clipping, and cleaned their backsides down. When clipping or cleaning, the sheep pens, which were usually cobbled, became slippery with droppings and water. I was a catcher for the clippers, and handling the sheep and the fleeces meant that you had to wash in hot water at night to get rid of the grease. In the morning, your bib and brace still smelt of the sheep. The shearers, using hand clippers and triangular wooden shearing stools, cut the fleece off in a single piece. The sides of the fleece were then folded in, it was rolled neatly up and the tail piece tucked round to keep it together.

The fleeces had to be gathered into wool bags, which were hessian sacks about six feet deep and one yard wide. These were hung from the rafters of the barn when not in use. It was usually a wet day job to pack the wool. The fleeces were rolled and tramped down: that is what my brother Jimmy was doing on the day war broke out in September 1939. The tops of the sacks were then sewn shut and taken away by wagon. One particularly horrible way of gathering wool was from a dead sheep: the carcase had to be dragged to a heap of field stones or to a bog for burial, but the wool had to be pulled from it first. The carcase was blue-green in colour, and it could be smelt up-wind. Since a horn or even the whole head of the animal could come off in your hands while it was being dragged, we usually worked them onto a sack, and then pulled that; the wool, at least, came off easily because of the long decay.

HORSES AND CARTS

Y OU have to imagine the roads in the Machars, not as they are now, but in the days when you could walk from Wigtown to Torhousemuir and never see or hear a car. In the thirties, very few farms in the Machars had tractors. I can remember the novelty of one at Newmilns before the war. Only later, in wartime, the government contractors could provide tractors from a depot at Newton Stewart to help with ploughing and harvest, so as to increase production, and it was only by the end of the war that, when walking, you might hope to get a lift from a passing car. Before that, the milk lorry was the main vehicle in which you could hitch a ride. Most farms, when we arrived at Torhousemuir, had perhaps five horses and some foals; at the

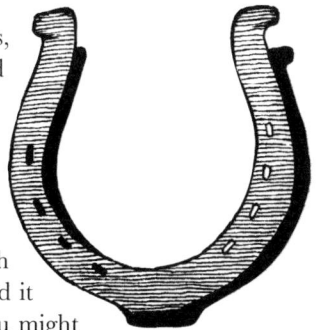

larger farms, there might have been cars, but even then they were not used very often. Even lorries were few and far between: there was one haulage contractor in Wigtown and another in Kirkcowan. It was only after the war that we began to see a few secondhand cars and that buses began running on two days a week between Wigtown and Kirkcowan.

We had a *half-bred*, which was a mixture of an ordinary horse and heavy pony, and one heavy pony. We acquired the pony from the outgoing tenant, but later had to replace it by buying one from a man in Wigtown. It had a personality of its own and would bite on occasion. The horse had been bought at the Newton Stewart horse-sale, held every October. Although we could have bought bigger horses, it had to be considered that a thorough-bred Clydesdale might eat as much as two bunches of straw in a night, as against one for a lighter horse; but by the time we went to Torhousekie, we could afford to keep two Clydesdales.

The horses were fed on crushed oats, and sometimes a bran mash, consisting of cut straw, bran and treacle. This was known to make their coats shine. Occasionally, they were also fed turnips. Every morning, the horses' tails had to be plaited and tied up round the runt of the tail, since this stopped them from sweating. For working days, we simply used twine, but ribbons would be used for shows. Harness was kept in good trim using boot polish. My father was particularly attentive to this, since he had won prizes when he was a ploughman for *grooming and harness,* and was later himself often a judge at ploughing matches for the special prizes in this category. For ordinary working days, we used the short collars, but high-top collars would be used for ploughing matches and shows.

To get a horse shod, we would have to lead it across the moor along a sheep-track, which was the right-of-way to the blacksmith at Causewayend. From Torhousemuir, the path we followed to Causewayend crossed Auchleand and then Cairnhouse moor; it joined a field road to Cairnhouse and then the route was down the main road to Carsegowan and on to the smiddy. The blacksmith was John Cain, of Whithorn. His shop was behind the house, which was semi-detached and shared with his neighbour, John Campbell, the joiner. I was probably about 14 or 15 when I began trips to his smiddy, perhaps once a fortnight, either with horses, or with coulters and plough socks.

The smithies – there were two others at Spital and at Malzie - were busy places, since they had to fit cart-wheel rims, and mend farm implements, or repair ploughshares needing extensions. It probably took about 3 hours for the smith to make and fit the shoe, and you could only hope that you would not arrive to find the smiddy full of other customers. My job was to hold the horse and help strip the old shoes. A cloth would be tied over the hoof to keep the feathers away from the foot. The blacksmith would try the shoe hot, then alter it according to his measurements, and finally cool it. There was a very distinctive smell of burning hoof as it was tried on hot.

Not every smith fitted shoes hot: I have known Bob Adair of Malzie carry four shoes, two under each arm, forced into the neck and armholes of his old jumper which had been holed many times by many sparks, with a tool bag on the front of his bike, and ride off to *cold-shod* a horse. That, however, was not common practice. By the time we lived at Torhousekie in 1945, Bob Adair was our blacksmith. He worked at night and you would often find two or three people gathered in the smiddy for the evenings. His father kept a tame raven, which occupied the smiddy and would croak: *Say hello, Joe.* We often had cause to go to Bob Adair's in the evenings, taking a ploughshare, and returning for it the following evening, since we always kept a spare coulter and sock in case of breakage. From him, I learnt to help him make the wedge-shaped groove in the shoe by holding the grooving tool which he hit with a heavy hammer. The groove was wedge-shaped so that the nails continued to grip as the shoe wore down. The ends of the horse nails were twisted off, and bent over with a specially designed tool. The horse's hoof was then dressed around the shoe.

Carts were the basis of all farm work; adapted in various ways, they would be used for hay and harvest, root crops, supplying peat, and carting the milk. The carts were built with wheels which sloped inwards, so that on a hard road, only half an inch, or an inch, would be bearing on the road. Presumably, when the cart was fully loaded, the wheels tightened and straightened up under the strain. The Wigtownshire carts differed from the Ayrshire carts, made by Jack of Maybole: the shafts of the carts in Wigtownshire were tapered to the horse's shape, but were straight in elevation, whereas Jack's shafts were shaped in elevation, but straight in plan. The carts were all made locally, and the making of a good cart was a matter of pride for the joiner.

The wheelwrights made the wheels for the carts. There was Rennie in Kirkinner and John Campbell at Causewayend, right next door, as was very common, to the smithy. The wheels were made in sections, known as *filets*, and a metal rim would be fixed round these, usually while it was still warm, so that it would tighten on the wheel as it cooled. In summer, the filets would dry out, leaving the iron rim loose, so we'd cover them with wet bags at night, and park the cart in water, to ensure that the wood swelled. Cart axles had to be greased, which entailed taking the axle half-way out, greasing the inside, and then pushing it through to the other side, to expose

Wheelwrights

that side of the axle; it was no compliment when we referred to the wartime margarine as *cart grease*. Even with proper maintenance, the iron-shod carts were hard to pull and rattled terribly. By the time we moved to Torhousekie, it was common to see wheels and axles taken from lorries and fitted to carts by the local blacksmith. We had two of these two-wheeled box carts, one of which was iron shod like this; the other had rubber tyres. Each cart had a number and a name-plate: one of ours, I remember, was labelled *D. Laird, No.3, Torhousekie*. The box-carts were used for hay and harvest. We also had a spring-cart for light work, such as taking the milk to the road-end.

The cart could be adapted for different purposes: an eight inch extension all round would deepen the cart for carrying turnips; these extensions were called *shelvings*. For harvest, slatted harvest- or *corn-rails* were fitted, which extended over the cart's sides and made it a foot wider on either side. The load would also extend over the horse's back. These extra pieces were stored in the cart-shed during the winter.

In the 'thirties, the bicycle was the universal method of transport. It was used to cycle to Wigtown on a Saturday night for the pictures, to the meetings at the road-ends, or to join up with friends our own age-group. We learnt to ride them in the fields, using an old lady's bicycle, which did not have tyres and so could only be used on the soft ground, nor a chain, so that we could only coast downhill. If I could manage a stolen moment with one of my brother's bicycles, when I was too small to reach the pedals, I would cycle with one leg through the bars, and one trouser leg thickly marked with grease; but in general, our elder brothers did not like to loan their new bicycles. When I was 15, I bought my first new bicycle for £3 10s, from Gibson's at Whithorn; the bicycle was inevitably the first item to be saved for and purchased by any boy who began working.

Given the limited means of transport, people had to show initiative in sharing bikes. When expecting a guest or a brother at Wigtown by train, you could, for example, ride one and push the other simultaneously. You could also offer a *lift on the bar o' the bike* or on the handle-bars: sometimes there were three on a bicycle. Girls could be given a lift home after a dance sitting side-saddle on the cross-bar. Sometimes, if a bike was to be shared between two, the arrangement was that you would cycle the length of two telegraph poles, leave the bike, and walk on, allowing the person on foot time to catch up, pick up the bike and overtake you on foot, repeating the process. It was nothing, in fact, to walk great distances in those days, and the roads were much busier with people on foot: during the day, we might walk on errands connected with the croft, and again at night, we would walk miles in groups or couples, to attend social events.

POACHING

Rabbits

Poaching was something which came naturally to us, and, as a small boy, I remember picking up techniques from my elder brothers; later, I remember hearing all the methods and tricks of the trade discussed at the Torhousemuir road-end, where poaching stories were exchanged.

There were three chief methods of catching rabbits, which were, in order of the legality and openness with which they were pursued: ferreting, snares and the long-net. Catching rabbits, I should say, was never pursued for killing's sake: it was often for a dynamo for your bike, or an extra night at the pictures. It was an unwritten rule that we did not ask our parents for extra money, and so we had to be resourceful in earning for ourselves.

Ferreting was usually carried out by daylight, generally on a Saturday afternoon or Sunday. My parents were well-aware of what we were doing, and sometimes it was a service which was welcomed by farmers or neighbours, where there were rabbits eating new corn or turnips. The ferrets would be hastily put inside our jacket pockets, or inside our shirts, and we would take off after lunch, usually on a Saturday afternoon. We kept the ferrets in the yard of the steading, in a box with a run and a sleeping compartment. They were usually fed on bread and milk, or, if you were gutting a rabbit, on some bit of rabbit offal.

The principle was to send the ferret down the burrow, and to have covered what we called the *scoot holes*, – the emergency exits which the rabbits had made from within, and which were often disguised with long grass - with purse-nets. The nets, which were known also as bag or *poke* nets, had about a three foot opening at the top, which was threaded with a draw-string. You put the net over the hole, spread it evenly round, and pegged it with a pin to hold it when the rabbit ran into it. You always had to have a second net ready in case one rabbit was trapped, and a second one quickly followed on its heels. One problem with this method was that ferrets could *lie up*, which is to say, simply come to a halt, either because he had come across a rabbit which would not budge, or for some other reason. He might lie still for as much as two hours, and then the only option was to dig him out, or simply wait for him to appear.

Purse-nets might be used without ferrets: there was a whinny knowe near Aaron Edmonds' croft which was a favourite place for rabbits. As they saw us coming, they would run for cover in the gorse. We would place nets on the entrance tunnels to rabbit warrens and my brother and Aaron Edmonds would tramp on the gorse to scare the rabbits out: as they ran back for cover in the burrows, they would be caught in the nets. They had to be killed with a sharp blow behind the ears, before they began kicking and potentially destroying the net.

The various methods of snaring or netting depend on the fact that rabbits are creatures of habit. They always follow their familiar runs. For snaring, we always set near a boundary, where we could see a path near a gate. Rabbit runs were different from hare-runs: these were wider and more distinct. For a rabbit snare, we set at a distance from the gate, because the rabbit would be moving too slowly under a gate to be snared. For hares, we would set close to a fence, because it would jump and hang itself on the snare. Auchleand Moor was a good place to snare rabbits: they had cover in the heather, and would come out to feed on the shorter grass. Snares would be set at night, and would have to be checked first thing in the morning. Usually, the rabbits would not hang themselves in the snare, but would have to be caught hold of, held up by the back legs and chopped smartly behind the ears.

My brothers no doubt showed us younger ones how to make the snares. Snare-wire, eyelets and brown twine had to be bought in. The twine had to be treated with something to discourage the rabbits from chewing it. Snare wire was brass and very pliable, and could be bent into any shape. A rabbit snare consisted of six strands of snare-wire, while a hare-snare had eight. To twist them together, you started with an eyelet with a nail through it; it was fixed to a hook on the handle of a bucket with stones in it. If you then spun the bucket, the wires were twisted into a single strong wire. The end was then passed through the eyelet, and the noose was then fixed to a

piece of snare cord using a reef-knot. The twine was then fixed to a wooden pin with a clove-hitch. The pin had a notch in it so that the twine did not slip out of it. The *teal pin*, as it was called, was set into the ground at an angle and was used to set the snare at the right height. It was made of hazel or rowan, and was about seven inches long. The height between the loop of snare wire and the ground was measured with the width of your hand. The effect of the knotting was that it was a running noose, and tightened as the rabbit ran into it. A hare-snare, on the other hand, would have a ten or eleven inch pin driven into the ground and the noose would be six or seven inches in diameter. An alternative when snaring hares was to tie the snare to a heavy stone. Occasionally, the white hares, which came down into the warmer coastal lands for the winter, especially if the winter was hard, would get snared. They were shorter than the ordinary hares, and we didn't really want to catch them. Because of their size, we would eat them rather than sell them on.

I was probably making snares from the age of about 7, and was ferreting from about the age of 12. You had to get up early before school to get round the snares, and check them again at night, or move them to a more promising site. On occasion, my dedicated pursuit of rabbits got me into trouble. I can remember one day in early autumn, when I was at Wigtown School, staring out of the window towards the tops of the tall trees across the road at Albion Cottage. The wind was blowing strong, and the undersides of the leaves were turning upwards, which was a sure sign of rain. I was thinking of my snares and where I should set them that night when I got home. During wet and windy weather, the rabbits run faster from their burrows and from shelter towards their feeding grounds, where the grass is shorter and the area is more exposed. There they can see and hear predators more easily than in the long grass and scrub being tossed by the wind. A fast-moving rabbit is more easily snared, and as I was speculating on all this, I felt a sudden sharp blow to the side of my head, and a loud "You're dreaming again, boy!". I had been so deep in my rabbiting plans, that I did not see or hear her approach from behind, much to the amusement of the class. I wasn't sorry to leave my school-days behind me.

Long-netting, by contrast, was something I came to when I was older; I probably began at about seventeen and was probably nineteen years old when I last long-netted. It was a technique used at night, though it depended on much the same principle as the snaring. This was largely carried out away from home, on neighbouring farms. There were boundaries which poachers observed, and we knew that towards Kirkchrist, which was a good area for rabbits, we would have got a hiding from other poachers if we had been found with a long-net. Rather than that, we settled for fewer rabbits elsewhere. We ourselves went as far as Barrachan Farm on the Mochrum Park estate: there were steep hills there, and the rabbits would run downhill into the nets. We were not beyond playing a trick on other poachers, if we could get away with it: on one occasion when the Rennie brothers and I were setting a long net, we discovered, near the Spittal, along the Bladnoch river bank, a pair of trousers neatly folded on the gatepost. We knew these must belong to a fellow poacher, but one who was after salmon, and that the trousers would be his respectable going-home pair. We obligingly dipped them in the river and went our way with the net.

The net which belonged to my brother Jimmy was one hundred yards long and was designed to be set down by a hedge. It required two men to set the net. The idea was that the rabbits would be scared from their feeding grounds towards the rough ground, and be trapped in the net. We would plan to make three sets in a night, in spots quite close together, and still be back home for the early milking. The catches varied from two to ten rabbits, but probably averaged about half a dozen. It could be more: the first time he set it, he caught fifty-four rabbits in three settings. It probably took about half an hour for a set and drive. You could, of course, be unlucky, and on one occasion, we returned without a rabbit. Fortunately, however, we spotted rabbits in a barn, which the keeper had caught, and took these in compensation for our bad luck that night.

The net was about three feet deep; they could be bought at Wigtown, but my brother Jimmy sent away to the *Scottish Farmer* for them, partly so as not to attract any notice in the town. The net had a top and bottom cord, running the whole length of the net, and a large iron pin at each end. When the net was to be set, the iron pin would be pushed into the ground, and as the net was paid out, we would insert a hazel stick every ten yards, leaving a bit of bag in the net. The cord would be twisted around the stick to hold it in place, and this pinned the bottom of the net close to the ground and the top about eighteen inches above ground level. One man carried the bag of hazel sticks on his back, and put pins in, while the other man paid the net out. Someone would then be responsible for scaring the rabbits into the net and another would kill them off once caught in the net. The driving was done as quietly as possible since it was night: it was enough to rattle a matchbox or tap your leg with a stick. The setting probably began at about 10 o'clock and we might finish after midnight.

When gathering the net back in, you pulled it back onto the metal pin in three foot folds, so that it would pay out easily for the next set. The right conditions for long-netting were moonlight and wind; a night which was black-dark made it difficult to see. We also had to avoid cattle straying into the net, and on one occasion, at Barrachan, a horse got caught in the net.

We would carry the rabbits we had caught in a hessian bag on our backs until we had finished for the night, and then gut them as soon as possible because we preferred to gut them warm. Our game bags were made of hessian sacks, doubled over at the top, with small stones in the top corners, round which cord would be tied to form a strap to wear over your shoulder. The guts were usually left lying on the ground, since the birds would eat them the next day. Occasionally, we might take our catch to a quarry and gut them there. We pulled the guts out of the rib-cage, and avoided bursting the stomach. You knew by feel when you had got behind the stomach. The liver was left inside the rabbit and the skins were left on them.

We would couple the gutted rabbits by first pushing the rabbit's paw between the bone and sinew of its other hind leg; then we would do the same with another rabbit, interlocking its back legs with the first one. The weight of a good catch of rabbits was something to be contended with if you had a long way to go. Pairs of rabbits could then be slung over the bars of a bicycle and sold to John Rennie in Wigtown, or Plunkett in Newton Stewart. We usually sold them within a day of catching them, hanging them in the barn until we could go out to sell them. Some of them were sold over the counter, but mostly they were shipped in crates away to the cities. Bradford was a place where you could get double the local price. No one bothered if you were seen to have rabbits for sale; catching them was an accepted thing to do, and they were regarded, in any case, as a menace.

Occasionally, we skinned our own rabbits, rather than leaving it to the game-dealers; that was if we were going to eat the meat ourselves. We could get a penny per skin from the visiting hawkers. To skin a rabbit, you start at the back-leg, slitting the skin from the crotch up the backleg towards the knuckle-joint, and cutting the skin at the paws. Then you would slit up the belly towards the front legs. Pulling the skin off the back-legs, you would pull off the tail part, and then take it off the carcase, rather like taking off a jumper inside out. Then you would feel down the front legs, and cut the skin at each paw.

There were some short-cuts to getting rabbits: the three or four cats we kept for ratting would occasionally bring home rabbits, which we took away from them and served for dinner. We also knew that stoats freeze rabbits, mesmerising them by running round them in a circle, surround them with their smell, before biting their necks. If we heard a squeal nearby, we would go out and get the rabbit. If we were too hasty, the stoat would take off, and the rabbit with it.

It should not be forgotten, either, that we always had a catapult in our inside jacket pockets. If you saw a thicket of grass round a briar, you would take aim with a stone, because you could almost guarantee then that a rabbit would be *clappit* inside it. Once you had stunned it, you could finish it off with a stick. I was once too optimistic and hit my target within a thicket, only to find that it was an old hen sitting on eggs. We made our catapults from ash-sticks, with a natural Y-shape. We would cut rubber from the inner tube of a car, and get the leather tongue from an old boot for the pouch. When I was older, I would also take aim with my brother's .22 rifle, which could be used against sitting targets, like rabbits in a *clap*.

Gate-netting

Usually, hares were snared or shot, but gate-netting was a seasonal method specifically for catching hares. It was practised after harvest, in September and October, when the hares were eating the fresh grass on the harvested fields. They were very plentiful in those days, and had not yet been affected by insecticides which they licked off their paws and which killed them. I began gate-netting at about the age of 17, without my parents' knowledge. Gate-netting was not like ferreting, which they approved.

I used to go with Pinty MacMillan and his old collie dog. *Old Pinty*, as we called him, though he was old only relative to us, was a well-known poacher, and lived in a cottage at The Grange farm. At the time, I was working on a farm with him and got to know him; he would send word up to me that he was going gate-netting that night and we would arrange to set off at about 9 o'clock and returning at about 1am. There was a certain anticipation during the day preceding a poaching expedition. We would go perhaps once or twice a week, catching four or five hares in a night. Later, I knew another keen gate-netter: Bob Adair, the smith at Malzie. He had a stag-hound he used for chasing the hares in, and he set a record on a hill at Boreland for the size of the hare he caught.

Pinty was known for other poaching feats. He shot pheasants out of the trees round Mochrum Park big house, since the keeper lived far away, round the *back o' the water* at Hillhead. He also poached salmon in the Bladnoch: he would locate the rocks and ledges where the salmon were during the day, and poach them at night, using torchlight and a barbed *gaff*, so that, once speared, the fish did not escape. Once he showed me how he could escape from the keeper by floating down the stream half-submerged. It was summer and we were all taking a dip: he showed us how he would float, showing only his nose and the curve of his belly above water, propelling himself downstream with his hands.

The nets we used were about eight foot by four foot deep with two strings at the corners. A net was hung upon a gate-post, with a little stone at the bottom to keep it open and one on each post to hold it secure, but not tight. The net could also be used at a hole in the wall or *lunky hole*. This is because hares will head for a gap or opening when running from danger. It was of great importance to have a good dog - one which knew how to keep the hare on the move. Pinty would set his dog off by a quiet: *On you go*, and he raced off into the stubble field. That dog never made a sound: this would have given the game away and lost us the hare. We once went with the Rennie boys whose dog yelped at the critical moment. Although later I made my own nets and used them on our home ground, I never had a good dog like Pinty's.

Two gates in a field meant that you had to have two people manning the nets. The technique was that you would sit beside the wall, and you could hear a thumping as the hare rushed through the stubble. When it hit the net, it would roll into a ball. You then had to break its neck, shake out the net, and replace it on the gateposts. I always remember being with Pinty's wife, Peggy, near the Standing Stones at Torhouse, by the river, on duty at a gate. She always wore a Home Guard coat when out poaching. She had the job of springing forward and breaking the hare's neck. It was important to kill the hare quickly, because if it kicked and struggled, it might have burst the net; it was also important to get the net back up again in case another hare followed.

I suppose I was about 13 or 14 when I learnt to make my own nets. Mrs. McGeoch's brother, who was crippled and lived at the Wards farm, near Whithorn, used to sell them for 7/6d, but I wanted to learn to make my own. The person who taught me was Tom Curran, a retired gamekeeper who lived at Duddingston Lodge, near Torhousekie Croft. You attached a hook to a wall, and got hold of a stick, which we called a *skemmel*[#] and which regulated the size of your mesh. For example, for a purse-net, you would use a skemmel about eight inches long by three-sixteenths thick and two inches wide; for a gate-net, on the other hand, you'd use one about fourteen inches long. They were usually taken from an orange box. The net cord was reddish-brown. You would probably take about a week to make a gate-net, working an hour or so each night, but a purse-net would only take about two hours. The net cord was threaded onto a needle (my first needle was wood, but it could have been bone) and attached to the hook or a nail; then you formed a loop round the skemmel to form a knot. Then the same again: turn it round the stick and form a knot; then slip it off the stick. The

Cf. note on the word below, page 81, Ed.

stick was as smooth as possible to make the slipping on and off of stitches as easy as possible. The whole process was repeated until you had eighteen loops on the skemmel, and then you would begin reducing by knitting two together. That would give you a diamond shape. To make a purse-net, you also added a draw-string, made with a soft cotton cord; it was threaded through all the loops and an end was left loose to tie to the pin.

Fox-hunts

About once a month in spring and summer, the word would go out that there was a fox-hunt on a Sunday. This was because most farm-workers had some leisure on a Sunday afternoon. By then, I was about 19 and had my own gun. A man from Kirkcowan, Bob Drysdale, had a couple of dogs. The largest and longest fox hunt I can remember began at Glaisnock, Kirkcowan, and a line of men would spread out past Barraer, through to past Merton Hall, and then going on to Barrhill farm on the Wigtown-to-Newton Stewart road. Not surprisingly, the shooters were tired at the end of the day, probably more so than the fox. I never actually successfully shot a fox, and I think if you did, you would have to be pretty close, or take him by surprise.

There must have been about fifteen or twenty guns, and a couple of greyhounds. The foxes would be lying asleep in the gorse; on being startled, they would either lie still, or get up and run. The dog fox would swing his tail round in a circle and urinate, which would set the dogs zigzagging after the scent, and give the foxes a chance to gain time and escape. I had one dubious success on that hunt: I shot a roe deer at the end of the day, which I found caught in a briar. My partner and I tied its legs on a stick and carried it home. Unfortunately, after we had cut its throat to bleed it, we ended up having to cut its head off completely, because it got caught on bushes and briars as we made our way home. We skinned it, cleaned it and cut it up to share it.

Game birds

I began shooting on my own account at the age of fourteen. Prior to that, my brothers would help me aim, or lean the gun on a dyke or fence. During the war years, every shot had to be a kill, because of the shortage of cartridges. We could aim at pheasants running down the drill in the turnip field. To get a clean bird, the aim was to hit the pheasant's head. After the shot was fired, you had to sit tight to avoid Gill the Keeper, who might be waiting to see who would pick up the bird. Often, we waited until lunchtime before we collected it.

According to the keeper, if you came across a pheasant caught in a snare by mistake, you were supposed to toss it up in the air, and see if it could still fly. If so, it was supposed to be returned to the wild. Of course, if we did find a live pheasant, we would stand on its head to make sure it couldn't. We also set snares deliberately for them: they could be easily snared after the potatoes had been lifted, between 5-6 pm. when they came to feed on the small potatoes. After feeding, they would draw their heads back up through the snares and pull the snares tight. Wild ducks were also attracted by the small potatoes. My parents would not have encouraged this, but if my father did not know the source of a meal, he certainly suspected that game-birds were served up as chicken: "Damn the chicken that that is", he would say. There were other methods which we did not try, but heard of, such as putting corn in a bottle to trap them, or impregnating corn with whisky to make them drunk.

On the whole, the partridges were left alone; some people did shoot them in the winter months for sport. They nested in the hayfield; usually 8 or 9 were to be found together. Wood-pigeons were usually too small to claim our attention: you needed four or five to make a meal. Occasionally, we would get a wild duck or goose. The geese were shot on the meadows near the river Bladnoch, mostly by my brother who was home from the Army and had a gun. They were generally hard to shoot because their feathers resisted the pellets; wild ducks were rather easier to kill successfully. The geese had to be plucked and singed. They had quite a strong fishy taste. We would get a tin of methylated spirits and light it to singe the down off. It was quite an effective method: a polish-tin lid full of methylated spirits would be enough to singe a hen and the meths. burnt clean.

Fishing

Fishing was never something we planned ahead or worked hard at: in summer, if we had nothing more pressing to do, we would wander the countryside, more often than not we would try to fish. Fish represented a change of menu, but it was not essential to our household. The burn near our croft ultimately became the Torhouse burn lower down. It was brown with peat from coming off the moss; at the time, there was no weedkiller, slurry or crop spraying, so that the water must have been remarkably pure. In various places, the stream was dammed for the cows, since we had no mains water to supply them.

If we went to the stream at the bottom of the field and followed it to near Ha' Hill, there were good places there to *guddle* for trout, wherever you could lie on the bank and feel around beneath it. I was probably 12 or 13

when I went regularly. In some places where the stream was partly dammed, we would look out for large stones in the bank, where the trout could hide. We would feel for the fish under the stones. You had to get them by the gills between your two finger knuckles, and grasp the tail with your other hand. Once secured, you would thrown them well onto the bank, and then strike their heads against a stone to kill them. They might be as small as six inches and we would rarely trap more than four. The risk you ran with guddling was that there could be huge eels lurking beneath the favoured stones, and these could bite. We had known them swallow hooks whole at Torhouse dam. The dam was reached by following the stream past Ha' Hill, skirting the *big house* and making for the bridge at Red Brae. The dam created a pool about two foot deep, and there you could fish for trout with a rod and line. We never attempted to fish the river by fly-fishing.

Another method was to use a snare or *girn*. This was made of a single strand of snare wire, formed into a loop of about two and a half inches in diameter. It was slipped over the trout's head, and then the trout was, in theory, thrown out onto the bank. In my experience, it was usually unsuccessful and required great patience.

Sometimes we went down Barallan, or the *thirty acre*, and fished the river Bladnoch for perch. No one worried much about our fishing for perch; we could obtain permission to fish and we could buy a hook and line in Wigtown. Others may have poached at night, probably using a net, but we were usually out in the afternoons, until about eight in the evening. The perch would be in deep water. We would be equipped with a six or eight foot length of hazel or rowan, a line and a jar of worms, which we had dug out of the midden at the croft. The float was set about seven feet up, allowing the bait to go down to where the fish were feeding. If they took the bait, the float would gently bob, and if you flicked it up out of the water, you could catch them. If we ran out of bait, we made use of tips learnt from the older boys and would cut the red part of the gill from a fish we had caught, or even use a red piece of cloth. The perch was good for eating, but it was scaly and had to be *ploated* in boiling water. Then the scales could be scraped off. At most we caught twenty perch, but usually about half a dozen. Your luck was in if you could get them at feeding time.

I have one less pleasant memory of Barallan. We had fished successfully, and had hidden about four fish each in our pockets. They must have fallen through the holes in our pockets into the jacket linings. When we emerged at Torhousemuir road-end, where the men had gathered at night to talk, we were questioned and denied having been up to anything in particular. Unfortunately, it was more than obvious there was something out of the ordinary about us: the fish had been spawning, and what we called the *milk* was running out of the hem of our jackets. As punishment, the men downed us on the roadside and *sichted* us, which is to say, pulled down our trousers in full view of all.

If the fish did not catch at Barallan, we would fish the pools between there and the stepping stones. If all else failed, we would wander up to the ruins of Torhouse mansion, often in search of pigeons, since the buildings were full of rock pigeons, and particularly of pigeons' eggs. Up there, it was so quiet that we found the wood pigeons built their nests low down and these could be easily robbed.

Wild foods

We would often go out to collect eggs from the wild: mostly, these were whaup eggs, peewheep or gull eggs. This we did for eating, although we did at one point have a collection of blown eggs.

We gathered the eggs from the nests: from a clutch of four, we would take only two eggs, and leave two to hatch. If you took gradually over a period of days, the hen would not forsake the nest. Whaups nested on a raised area of ground in marshy, rushy land or on the moors. We would watch the whaup circling and calling; she would descend steeply, and land within five or ten yards of her nest. We could be as far away as three to four hundred yards, but we would watch her walk towards the nest, then see her taking off again and note its exact location. Once the birds were alarmed by our approach, peewheeps, like partridges, snipe and plovers, would try to distract us from the nest by running with their wing trailing as if injured. They would then take off and fly in a wide circle, to return to the nest behind us.

Since the birds laid each day, we would see if she had one egg, and then we would leave it. We would return to take the second, then the third, and leave the fourth. The eggs were rather bigger than a hen's egg. The point of waiting for a day was to guarantee that the egg was not *caught*, or already developed into a chick. There was a way of testing for this: if you put the eggs in cold water, if they floated, they could be *clockin'*, but if they sank they were fresh.

The gull eggs we took were those of the black-headed gull. We had to walk to Barvennan moor to get these. We used gull and whaup eggs for cooking to eat ourselves, but others collected them for shipping to London, where they were considered a delicacy.

The neighbouring farm was densely populated by peewheeps. There were hundreds of them, and you could barely walk without treading on the nests, which consisted of small scrapes with some wisps of straw. The eggs were green with a black speckle. If the eggs were fresh and the hen was still laying, we could take them all, since the bird would continue to lay. In sowing time, or when harrowing or rolling, it was common to come across a nest on the ground: some might be put inside the sower's cap, or the nest might be carefully moved away from the machinery, and then replaced where it had been, with most of its contents in tact. Water-hens laid up to ten eggs, in open, wet ground. The *rashy field* was full of plovers' nests, but they were too small to be worth cooking. The cooking technique was simple: we would light a fire, get a billy-can and hard-boil the eggs, eating them out of our hands. Generally, we would go nest-hunting when we had nothing better to do.

We grew up with an understanding of animals, birds and plants. We would watch stoats and other animals, and – though we were boys - collect wild flowers, like lily of the valley or scented stocks, to press them in a book, and gather *fairy tatties* in May and June. I can remember sitting on a bank by Duncan MacGregor's, which was a mass of the flowers. The tattie was about the size of a plum. We would rub the skin off with our fingers, wipe them on our trousers, and savour the nut-like flavour. They were also known as pignuts or earthnuts. There were always tricks to stave off the worst of hunger and thirst: *sourrocks* were known to draw the saliva, and we would chew on the leaves if we were thirsty. The leaf looks like a small dockan. A single stem of honeysuckle would contain a whiff of nectar, if you sucked the narrow end. There were raspberries in the wild, if you knew where to go, crab apples for jelly, and, in season, we would gather enough wild mushrooms, particularly where there was plentiful horse-dung, to be fried or made into ketchup. Perhaps the one animal we gave a wide berth to was the adder. They would appear in the yard at Mossend. The cats would soon alert you to their presence by hissing and arching their backs.

PEAT

OUR peat came from the moss: each croft was granted a *breast* of peat to cut, which was included in the lease. The location of the peat-breasts was about three-quarters of a mile from the croft. Each breast, which was about ten or fifteen yards long, was allocated to a specific croft. Beyond the area allocated to Mossend, there was a ditch, then a breast for Hillview and one for Balmeg beyond. Behind us was the breast for the farm of Knockmore; since they had given up using peat, their breast was left about thirty yards behind ours. The Edmonds, Harpers and Forlows were the only other families who used the peat breasts during our time at Torhousemuir. The larger farms had all begun to use coal, but we calculated that the peat was free, and that if you had plenty of labour – in our case, three children at home – it saved hard-earned cash, which might otherwise have had to

go to the coalman. By the time we left in 1945, we were the only family left using the moss, since the incoming tenants did not cut peat.

We cut our peats at breast-height, not underfoot. You worked at a level which was about seven foot below the ground level. Generally, it was my father's work to cut the peats, but I was about ten when I began helping with lifting and carting. The cutting began with lifting the sod off the top; these were cut in sections of about a foot deep, and were full of heather roots and weeds. To do this, we used a *lurgan* spade, which had an exceptionally broad blade and a large cross-piece handle. It was also occasionally used in ditching. The turfs were carefully placed in the existing ditch, and were used to stand on and to keep your feet out of the water which constantly eased out of the peat. The first layers of peats, with a quantity of roots in them, were known as *hairy peats*, which retained a lot of water and took longer to dry out; below that, you cut into the thick dark layer, which was much harder when dried. The full length of the peat-breast was known as a *skemmel**. There was a certain rhythm to cutting peats, which involved moving as little as possible: my father would shove the peat-spade about two foot six into the peat, and simply twist round from where he was standing to off-load it onto a barrow. Undue movement would have meant that water formed around your feet.

We always worked with two barrows, or three if possible, so that the cutter was not held up while the empty barrow returned. The barrow was specifically designed for peat: they had long *trams* which went all the way to the front of the barrow, a plank bottom and no sides, so that you could tip the heavy load of peats off to the side. The wheel was wooden with an iron rim, or was solid iron. The pneumatic wheel, when it came, was a boon to the peat-cutter, since the tyre no longer cut into the moss, as the old iron-shod barrow wheels did. Usually, the barrow would be equipped with an old motorbike wheel. We probably began with joiner-made barrows, but later made our own.

Generally, the peat-cutting commenced in late spring. The peat spade had a face about seven inches wide, and went to a point; it had a lug on the side to cut the peats up the side. Part of the shaft was flattened and at least as wide as the blade, so that the peat could lie along it. The peats we cut were just over six inches square and two foot six long. Six or seven formed a load for an adult to push, but probably about five for a child. My mother, or myself and my brothers wheeled the barrow away from the cutting and tipped it onto a level area. After about a week, they would be spread out, and then turned. They were then stood on edge, or *fitted*, once they were relatively dry. The *fittin's* consisted of three peats in a pyramid,

---

* From the Scottish National Dictionary - SKEMMEL Also -le, -il, skem(b)le, scamble 1) In plural: a shambles, slaughterhouse; a meat-,arkey, orig. the tables or benches on which the meat was exposed for sale. 2) A peat-bank, the Hagg or area left in a moss from which peats have been cut, probably from the similarity to a series of benches. (Ayr 1928; Wigtown 1958). [Concise Scots Dictionary gives date of usage as 20th century] Ed.

with a space in the middle to let the wind dry them out thoroughly. When dry, they might be reduced in size to about 3 inches square, and between eighteen inches to two foot long. Today, the drying process has been eliminated, since mechanised peat-cutting involves compressing the peats, which squeezes out the excess water.

After some weeks, the peats would be barrowed to where a horse and cart could pick them up, because the moss was too boggy to support either. The barrows used for this were the ordinary dung barrows: the peats were now considerably lighter, and *shelvings* could be put on the sides of the barrow, to increase the load it could take. They were roughly stacked, and after two to three weeks, the peats were carted home, usually in late summer, after harvest. They were built into a stack, somewhere in the yard by the house. The stack had a specific structure, designed to shed water. The peats were stacked at an angle on the outside, rather like overlapping slates and brought to a point at the top. The whole stack, which had to last a winter, was about six foot wide, by seven foot high and ten or twelve feet long. The top would be covered with sacking, or a tarpaulin, with bricks on ropes tied at the corners to weight it down. The stack always outlasted the winter, and occasionally we had some left over in spring.

The peat-cutting was fitted in between turnip cultivation, hay and harvest. You might cut peat continuously for several days. The peat-cutting was usually carried out barefoot. The plague of midges on the moss was such that we could not carry out the work in the evenings. As a small child, I can remember plaintively asking my mother: *What do they eat when we're not here?* During the day, when the sun was hot and a good wind blowing, the conditions were best. There was also the hazard of the occasional adder, basking on the hot days.

The peat fire was the only source of heat and of cooking fuel. We kept a galvanised bath full of it beneath the kitchen table. When reduced to ashes in the hearth, it also liberally sprinkled everything with a fine dust. Your clothes smelt of it; the socks which hung on a wire over the mantelpiece were impregnated with it. If the peat was smoky, the smoke would curl over the edge of the girdle and affect the scones, which were then said to be *peat-reekit.*

School

SCHOOL probably had much less impact on us than our life at the croft and like most country boys, we were not interested in our lessons. Country children could easily be told apart from the town children: for example, they always wore heavy boots, rather than shoes. Generally, our home life was kept quite separate from the life we lived at school with other children, and, at home, the examples we followed came from our elder brothers rather than from contemporaries. The crofting life probably taught us to be more resourceful than our classmates: we may not have been street-wise, but we were country-wise and capable of doing responsible jobs at a very young age. We knew that we were different from the Wigtown boys, and that our leisure time was different from theirs: we always had things to do at home – calves to feed, animals to groom, peats to fetch. Other than farm work, we did have some activities indoors at home: we might learn drawing by tracing faces from magazines, which we had had handed on to us when they were perhaps a year out of date. Brooke Bond tea had an inner wrapper which could be saved and used as tracing paper. Reading had to be done in the living room, since the lamps in the bedrooms were too dim. Probably, however, most of my childhood was spent outside and most of what we learnt was from our parents, brothers or simply from observation, rather than from books.

Between the ages of six and fourteen, I attended Wigtown School, until I had to take on full-time work at home, owing to my father having an operation for ulcers. Being more than three miles from the school, we were entitled to a car to take us there; those who had the misfortune to live at a distance just under three miles had to walk. Wigtown school occupied the buildings it does now, without the extensions it now has; it also had the gas-works at the bottom of the playground, and the school garden, where we worked twice a week. Until we were in the Qualifying Class, we had one teacher, but after entering the Higher Grade, there were different teachers for different subjects. Some pupils might have gone on to the Douglas Ewart

High School, but most did not. For some, it was a consideration in the home whether the uniform was affordable; others were needed at home to make another wage. A lot of boys were simply passing their time until they could leave to get a job; in those days, there were always jobs on farms and the prospect of a wage was a real incentive to work. Even those three children I remember who were unable to cope with ordinary lessons took on jobs after they left school. Later in life, I thought perhaps I had missed out on further education, but most parents before the War did not push their children, but had to consider how they might earn to support the family.

On the whole, the education provided was fairly basic: most of my school years were war years, and, owing to a shortage of teachers, our curriculum was patchy at best. There seemed to be little science and no woodwork or practical training, because the male teachers had left for the forces. Some male teachers arrived with the evacuees, but only one stayed. The headteacher, the formidable Miss Esslemont, took over when the headmaster, who had been a Captain in the Territorials, left. Probably the evacuees boosted numbers, but I can remember twenty or thirty being in a class. There were quite distinct groups of evacuees, some coming from poor homes and asking for *the runt o' your apple* in the morning break; those from Jordanhill arrived in new brown and yellow uniforms. This set them apart, because very few of the Wigtown children wore uniform owing to the expense. There were also the *private evacuees*, whose parents paid for their board, or who were boarding with relatives. Mostly, they were put up in local farm-houses where there was plenty of room, rather than in farm-workers' cottages where the families were already large. A few stayed on after the war, perhaps working on the farms where they had stayed; others returned on holiday.

We country children took sandwiches for lunch, which could be eaten in an open fronted shed in the playground, with nowhere to sit down; the town children went home, since no lunches were provided by the school. I remember that some children had only treacle pieces for their lunch. In Wigtown itself, if we strolled into the town occasionally, there was soup or Bovril for sale in the lunch hour, and later on, I went to a friend's house for lunch, where I got a cup of tea. My parents gave the family a sack of potatoes in exchange for this favour.

HOUSEHOLD CHORES

MY parents had to pull together to keep the croft going. On the whole, my mother's work was concentrated round the house and the steading, but during hay and harvest, she worked in the fields alongside my father. The concept of leisure time as we know it did not really exist: there was little time off, even in winter. Perhaps a ploughing match, a farm sale or a funeral might have given the opportunity for an afternoon off.

All the cooking was done on the open fire. This was about three feet wide when we arrived at the croft, and my father reduced it to about eighteen inches wide, to form two hobs on either side. One of my mother's first investments was a cast iron pot, or oven pot with a carrying handle. It was probably about fourteen inches wide and seven inches deep. In this she did almost all the cooking. This hung on a chain, and it could be adjusted up or down a link to give it the correct distance from the fire. It was some years before we got a Valor paraffin stove and a double skinned oven which could be sat on the top of it, absorbing the heat through the bottom and up the sides. The oven pot, however, was remarkably versatile, when used by an expert. To cook scones, my mother would put a red hot peat upon the lid to heat the whole pot through; meat could be cooked by putting in a little water with it, and letting it roast with the lid on.

I still remember feeling the heat off the fire, changing when my mother moved in front of it as she cooked, and the smell of the peat smoke. This was often in the evenings, because if my mother had had no time to bake during the day, she would start up after seven at night. Apart from the oven pot, there was also the tattie pot, the porridge pot and the hen's pot. They would be laid aside after washing, sitting upside down on a shelf. These were all black enamel pots, grey inside. All the cooking had to be done and all the cooking utensils stored in the cramped space of the living room; sometimes a ham might hang from the ceiling, and would be used for slicing for the breakfast bacon.

Much, though not all our food, came from the croft itself, or from by-products of what we sold. For example, when a cow calved, the milk was too rich and yellow to be sent to the creamery. It was then used for making pancakes, and these were known as *beasning* pancakes. The milk acted as a laxative, because it was so rich. Drowned sheep, which had strayed into the moss, were given us free of charge. They were edible, because you could still bleed them. After a sheep had been hung up and bled, my father would skin and clean it, and then butcher it in the barn. In these days before refrigeration, the meat was kept in a zinc box with a perforated gauze door to keep the insects out, or wrapped and stored. Out of a butchered sheep, we would get pot-roasted or boiled mutton, and sheep's head broth, of which we grew heartily sick. The fleece would be sold to the scrap-merchants, and might fetch perhaps five shillings.

Wild game, such as hare, was boiled before being fried: the boiling took away some of the wild taste. Rabbit was stewed, or sometimes boiled and fried. The small of the back and the back legs were the best parts, better than the rib-cage or front legs. It could also be made into pies.

My mother reared the chicks and looked after the hens. Their dung would be put on the garden. The older hens would have their necks wrung, and would then be boiled; the feathers would be set aside for pillows. During the war, my mother won a white Wyandotte cockerel in a raffle and called it Churchill. It chased her up the yard one day and burst a varicose vein. She never suffered from that varicose vein afterwards.

Clothes were washed on a Monday, because the weekend was the time when each member of the family changed his clothes. Before modern notions of hygiene came in and modern fabrics, clothes were used for longer than now; when the collars were dirty, it meant there was time for a change. Clean clothes would be ready and waiting for my brothers to pick up on the following Sunday when they returned with their week's laundry. A rubbing board was kept within the galvanised bath in the wash-house. After scrubbing, everything was rinsed in cold water and wrung out by hand. Later we did acquire a wringer, which had two small rubber rollers. White clothes would be boiled in the boiler and *blueing* was used to whiten them. Blankets would be tramped in a bath to clean them. My mother patched and darned to make things last, and I remember her being able to knit with her eyes shut.

The iron was a box-iron, with a bolt which was put into the fire, then inserted into the iron and the hinged lid closed. Because there were no artificial fibres, everything was ironed, from towels to table cloths, working and dress shirts. As we grew older, we pressed our own flannels to a sharp crease, taking our cue from our elder brother who was in the Army.

The floors were washed twice a week, the lobby flagstones scrubbed and the hobs black-leaded. The living room had a steel fender with an ornate top, and the ash-pit cover was highly polished with black-lead. When the cleaning had been done, in the bedrooms and on the doorstep, my mother would use a piece of red tile or stucco, dampened and formed into a ball, to create a design on the stone. Quite often, flagstones round the house were also decorated. Sometimes, it was just the edges of the steps or hearths, but I can remember that at Torhouse, Mrs. Cannon drew a full picture of a large swan, sailing with its feathers ruffled out. By the end of the war, this practice was dying out, and painted steps with liquid linoleum in red or green came in.

My mother made all the rugs for the hearths and bedsides. To make a hooked rug, my mother would start with a hessian sack, and cut strips of cloth of different colours about an inch wide; these came from old dresses and pieces of worn out clothing. She always had a pattern in mind before she started. She used a clothes peg with a hook to pull loops of cloth through to about an inch and a half. The loops were then clipped to make them all the same length and to thicken out the pile. She began in the middle, worked up a diamond shape, and then went on to work the corners. To finish it, she would sow another sack on the bottom. In the bedrooms, we had linoleum in the centre of the floor, and the surround was varnished.

CLOTHING

N OTHING was wasted on the croft. We survived because things were re-used and made to last, and anything which could be made at home was made, in order to conserve cash. Our school trousers were made out of an old coat, and they would be lined with flour bags on which you could still read the lettering. My mother's aprons were also made out of a hessian sack. As children, we had cast-offs from my cousins, since their father – my uncle – was a draper in London. We therefore always ended up wearing their school colours to Wigtown School. Elegance was not in question, however; I have a photograph of myself wearing my V-neck jumper *backside-foremost* to protect my chest from the cold in winter. Old jumpers were unravelled to make into new socks, and socks, when worn out, were cut off above the heel and *re-footed* skilfully by my mother, using wool from discarded socks and jumpers. The result was often colourful. To help her with all her mending, my sister Minnie bought my mother a hand-sewing machine, and eventually a treadle machine.

Children might wear shoes designed to fit them at the end of a year's growth. In winter, we would add handfuls of straw to the inside of our boots to keep our feet warm. Hems were generous and meant for letting down. We, in our *combinations*, were luckier than some who had no underwear. I remember that many sported the *glazed sleeves* at school, which were the constant result of nose-wiping, and that this was particularly so among the evacuees. We had what we called a *donkey fringe*, and the rest of the hair was cut short all over. My father would do the cutting, using a pair of clippers. Wartime children had their hair regularly combed for nits, and Derbac soap was used as a lice-repellent. In those days, *going in to long trousers* was an event which marked a stage in life: we were as old as 14 or 15 before we got our first pairs. Before that, we would wear long shorts. Working clothes might be of any size and vintage. As boys, we wore men's caps and would put up with the jest: "If you didnae have ears, you'd be blind".

On the whole, colours were drab and fashions stayed for a long time. When the war came, clothing was rationed and remained on coupons until after the war. The usual dress for older men was breeches with long-johns beneath, a heavy *semmit* and a thick shirt without a collar. The long johns might be worn all year round. My father wore a grey shirt with a red or white stripe, and the detachable collar. On top of that would be a waistcoat and a tweed jacket and cap. It was only later on that the *bib and brace* came in, and a jumper was worn instead of a waistcoat. In many cases, the sign that a farmer was going out was the simple addition of the starched collar; the shirt was not changed. Likewise, the shirts or heavy semmits worn during the day might double as nightwear.

My mother wore a fairly standard working dress while on the croft. For milking, she wore a bag-apron and a kerchief or dust-cap round her head, because she leant against the cow's flank. In the harvest-field, or on the moss, she would wear a *mutch*, which she made with a four-inch peak reinforced with cardboard. This kept the sun off her face. She wore Lisle stockings, or sometimes fine woollen stockings, stays or corsets, clogs, a heavy skirt, a jumper, and a wraparound *pinny*. Her working skirts were made out of old coats, which supplied the heavy material she needed. For going out, she had a frock and a decent coat and court shoes, for riding on a bicycle to Wigtown. Once, I believe she went as far as the Edinburgh exhibition.

Ploughmen, who had a status above the ordinary worker on the farm, wore breeches with block-leggings. These were designed to keep off the mud, and were made of leather. They buckled on, using a pin which entered a socket, and a strap and eyelet which fastened the top. Boots and leggings were a matter of pride for some ploughmen and farmers. A decent suit and *fine boots* were essential for market days; the boots would have plain leather soles, rather than *tackets* and would be made of brown leather. Shoes were very little worn, particularly by the older generation. We might only have had one coat "for kirk, town and market", but it was essential at least to have one respectable outfit. For some, dressing up simply meant putting a clean shirt on over an old one.

Our own boots were cleaned after each use and put on shelves above the door. These were the school boots, but we also had older boots for working and heavy Wellingtons, which were left in the lobby under our coats. All the boots had metal toe-plates, metal heel plates and rows of tackets along the foot. They kept the leather off the rough ground, and, when worn to school, made recess a noisy affair. I was probably at least 13 before I wore shoes to school. If the tackets wore down, you could buy new studs and re-stud them yourself; ex-army boots could be made into work boots by hammering in new studs. It was a time at which most people had a boot-last at home. We also used bits of old harness to repair the heels. New boots could be bought from Stewart's in Wigtown, and *lea-rig* boots and leggings were advertised in the *Scottish Farmer*. Lea-rig boots were made of stout leather with turned up toes, so that you seemed to rock forward as you walked. On a wet day, school boots sometimes had to be slung round our necks and an old raincoat covered our heads, as we ran to the road-end in our Wellingtons. There, we would take them off and turn them upside down behind a dyke, and pick them and the old coat up at the end of the day. In our leisure time, we might go barefoot, and I can remember the dust running through my toes like water, in the long dry summers we seemed to have then. We would be forced by my mother to wash our feet in the burn or to pump water onto them before going to bed. Some children, from the burgh of Wigtown itself, came barefoot to school, and one family of boys were made to wash the cow dung off their feet before they came into class. They were sons of a dairyman at Carsegowan, and the teacher probably failed to appreciate that they had been at work long before they had walked the four miles to Wigtown and arrived at the school.

The War changed a good deal, including people's appearance, since it made available new materials and ex-Army supplies, and made other fabrics hard to get. The sharp appearance of service personnel also made a stark contrast to the usually un-pressed state of our country clothing. During the war, when money and supplies were short, ploughmen would cut the feet off Wellington boots to use them as leggings, and there were also re-conditioned Army boots to be had. When I was between the ages of sixteen and eighteen and working at Torhousekie, I wore ex-despatch-rider's breeches, which could be bought at Gowan's in Castle Douglas, a khaki or blue drill shirt, and leggings. I remember that the cloth rubbing on your thighs as you ploughed or worked in the cold and wet made them what we called *pisket* or raw. We also used ex-army gas capes to make trousers out of water-proof material, but it cracked and tore easily. My sister made my brother and me bathing suits out of black-out material.

Social life

THERE is little doubt that people were more isolated and less mobile than they are now. Torhousemuir, given its location off the main road, was more isolated than most. Quite often, the older women did not leave the estate at all, or, if they did, perhaps only once or twice a year. Their sons perhaps did the shopping by bicycle, and their husbands went without them into town, although even then, it might only be once in two or three months. Even the post might only come to your door once or twice a week, or it might sometimes be only once a fortnight. My father went out very little, but he did occasionally go to the market at Newton Stewart on a Friday.

Another factor which made life in the countryside vastly different then from what it is today was the fact that all the farm cottages were occupied, and there were large families in every croft and cottage. There were regular but informal meetings of the men in different parts of the parish: on a Tuesday night, men would gather at Torhousemuir road-end, and also on a Sunday. A few girls would also come, either farmers' or farm-workers' daughters. The Torhousemill field was used by the football team on a Tuesday, which entered the Summer League, and played against Whauphill and the Grange. These were mainly single men and the meetings were at their peak before the War, when many volunteered and left the area. At those meetings, the air was thick with smoke of Woodbines, Players and Senior Service cigarettes. Every man would have a cap, always worn at a jaunty angle: pushed forward, pushed to the back of the head, or else worn radically to one side. When they played football, most had no football boots and played wearing the tackety boots they wore for work. This was an occasion on which gossip was exchanged from different ends of the parish. Sometimes, the meeting would be at Culshabbin, and there would be no football, but the group would just sit on the road-side and talk.

Word of mouth was the universal method of spreading and obtaining news. The butcher, the baker, the grocer, the milk-lorry driver and the postman were all sources of information about the outlying places, and they were expected to carry the news. The postman, who carried our post until 1945, would always have lunch wherever he ended up at lunchtime, and almost more important than the letters and newspapers he might bring, he carried news and messages. There were a great many mobile services, since the population was much more static. There were two grocers and two bakers, from Creetown and Kirkcowan, which came to our road-end a mile away. We would buy four to six loaves to last a week. Paraffin cans would be carried by the grocer on a rack along the side of the van. We would either carry the cans, or transport them on the back of a bicycle.

As boys, we would be quizzed by neighbours as we passed as to what we were doing and where we were going, and these were matters to be thought over and passed on. Within the estate, neighbours visited each other; in winter, they would come with a byre-lamp. Of course, you could never neglect to stand and talk if you passed a neighbour, whatever pressing business you had.

Wigtown was the place to meet friends on a Saturday night, for going to the pictures; in general, Saturday was the night for going out since lunchtime that day was pay-day on the farms and most had the afternoon off. Three or four of us from the estate might join up and walk from Torhousemuir to Wigtown, a distance of over four miles. Noah Henry had the picture house there, which was an old tin shed, with a frontage which had been part of the Edinburgh 1936 exhibition. He also owned the picture house in Whithorn. In those days, there might have been three or four different pictures per week, the cheapest seats being threepence, and the most expensive two shillings.

The biggest holiday of the year was the Wigtown Show, and the first I can remember would be in about 1939. There were fruit stalls and trade stalls. Mostly, we walked in, to save on the bus fare, leaving our door – as usual – unlocked behind us. The countryside was totally deserted on Show day, and for many it was the only annual holiday. Like many others, however, for us the work of the croft carried on and we had to be back in time for the milking.

I probably began going out and about to dances once war had begun, and there were plenty of events organised to support the war effort. The Women's Rurals would organise basket-teas, dances, or whist-drives, and knitting-bees to make gloves, balaclavas and socks; my mother was a member of the Grange Rural. There were dances at Malzie or Dirnow: the country schools were where the Rurals met, and where weddings and receptions were held. Whithorn was also a favourite place to go, and dances were held at the New Town Hall. People in general travelled great distances to attend events: it was nothing to cycle from Torhousemuir to Culshabbin for a basket tea. It was four and a half miles to the Grange School, but old and young would turn up. Three or four women would walk together and the groups would swell as they walked: my mother would start out with Mrs. Martin at Torhouse Cottage. Occasionally, there were less conventional means of getting a lift: I remember two local girls sitting on top of dead cows on Dundas' Knackery van, on their way to a dance at Kirkcowan.

At the dances, someone might play an accordion or a fiddle, who had learned it after a manner, and there were eightsome reels, military two-steps and waltzes, *Drops of Brandy* and the *Pride of Erin*. We never really learnt to cope with the quick-step, though Kate McGeoch at Kirkcowan and Bob Phillips at Wigtown gave dancing classes, which cost 7/6 for six weeks. I went when I was 15 or 16 and there were twenty in the class. The bicycles were kept in the school sheds during the dance, and the first one in was the last to leave, since it took an age to disentangle the bicycles from the mass. We'd have prepared for a dance by washing with fairy or carbolic soap in a basin in the wash-house with a drop of hot water in it to take the chill off. For hair-washing, we'd use what was called *black soap*, which was actually a soft dark green soap. Women would wash more privately in the bedrooms. Even so, some of the farmers' daughters smelt strongly of the byre or of sweat. They had men's jobs on the farms, and could easily swing a man off his feet during an eightsome reel. Some of the men might dress up for a dance by donning a new bib and brace.

People would take time to talk before leaving, and couples paired off. If you were at home in time to feed the horse the next morning, no one minded much where you'd been. I can remember coming home from dances at midnight, when, with double summer time, it was still almost light. The mist was settling down on the harvest fields, and the stooks loomed above it. You could see the grouse settling down on the stooks for the night and partridges on the ground, amongst the stubble. You were probably chilly by then, and still had two miles to go.

In summer, we'd cycle from Torhousemuir past Malzie and into Whauphill, along the beech hedges, and down into Monreith. Usually, it was a Sunday, and we'd bathe at St. Medan's Back Bay. It was only hunger which drove us home. Between Mochrum and Barrachan one day, we stole turnips to stave off the hunger pangs, and were chased after by the farmer. It was hardly worth it: the turnips were immature and nippy. Sometimes, we'd go to the Black Rock sands, and it was not unusual for crowds to gather at the top, above the steep descent. A couple of Italian prisoners from a nearby farm would bring an accordion and sing, and large numbers of people would turn out from Monreith.

Funerals and marriages were events which made a rare break in the croft routine. For funerals, people wore heavy top-coats and bowler hats, with black suits. Women never attended. The ministers said the services at the door of the house, if there were a large crowd, since the coffins were always kept at home until burial. Teas were held afterwards, often at the house. If the house were very remote, the service was held at the graveside only. Marriages were often held in the manse, or at a hall. There were also receptions at the Creamery Hall at Bladnoch, where there were dances every Saturday night. There were always *blackenings* before a marriage: the groom, if the blackening party could catch him, was covered from head to foot in boot polish or anything else which came to hand. We were not church-goers ourselves and I don't recall many others from Torhousemuir attending church.

The war altered a good deal in our social life, as in all else. There were soldiers and airmen at Baldoon camp, and women began going out and about, into pubs, something which had not been heard of before. It would have been rare to see a woman smoking or drinking before that time. The ATS and WAAF's would go to Baldoon at weekends. Servicemen would attend local dances to inspect the local talent. That brought about occasional fights, when soldiers coming home fought with soldiers billeted here, or with the Italian collaborators. In fact, there was a good deal of discord leading to broken marriages during the war.

In 1939, *Wigtown Ploughman* was published by John McNeillie. My parents were readers of the newspaper in which it was serialised and there was a great deal of talk about the publication. The people I knew could well understand the brutality featured in the book, though others disapproved, because they did not wish themselves to be seen in that light. I was too young to read much of it, though I had heard of the places mentioned in the book. As yet, I had not been to Malzie, but we could see it from the top of the hill. In later years, Ian Niall sent me a message, when I was working in North Wales, to come over and visit him, and I always regretted that I did not make the time to meet him then.

Many of the things John McNeillie described were instantly recognisable. There was a high rate of illegitimacy: I can remember two brothers who fathered thirteen children between them, and only acknowledged one. Families were of course generally large: there might be anything from six to eleven children. In seasonal work, such as turnip thinning, of course, this was an advantage, since many hands could raise additional income. For some reason, the area round Whauphill was notorious for illegitimacy. Just as McNeillie describes, the sexual encounters I witnessed as a young boy were brutal and surprisingly public: it was sheer lust, no passion. The women were as coarse as the men, or at least accepted the coarseness. It was accepted that, on a Saturday, country girls and men would vanish together on the way home from a dance. Illegitimate people therefore referred to themselves as *weekenders* because intercourse was the aim of this Saturday night socialising. Wartime seemed to accentuate the problem: some married women, whose husbands were away in the services, had affairs with visiting servicemen or POW's, some openly, others very discreetly. Some, perhaps, were not flush with money, and this was a way of obtaining extra rations or money. By this time, however, perhaps the influence of Hollywood cinema made courtship a more romantic affair. Women, in any case, became more independent, as I said, smoking openly and visiting pubs: in that way, the war marked a complete change in the way of life, particularly for women.

SOCIAL ORGANISATION

AT first, our household was supported by the contributions of my three older brothers, who worked away from the croft during the day, but came home at night. John started work at Knockmore, and Alastair at Meadowbank. Later on, the two boys lived away and only came home at weekends. It was common for families to send sons away to bring money back into the home; working away brought in hard cash. In return, my mother bought clothing for them, mended and washed. My brother Jimmy continued to support the family home substantially after he worked away: his pay was 10 shillings per week, since 10 shillings of his pay was kept back for his keep at Clauchrie. He gave nine shillings of the remainder to my mother to keep up the family home, and kept one shilling for himself, to pay for the pictures on a Saturday night, and for a packet of chips. He said that he never felt the desire to spend more than that, and even though he did not eat at home, he felt obliged to support the family with the majority of his wages. He came home on a Sunday and went back at night in time for the milking. My brother Alastair, however, drank and smoked and had little left at the end of his six months, after *subbing* his pay for the preceding half year. My brother John was much steadier: he would sacrifice going to the pictures in Wigtown if he had had to buy new batteries for his bicycle-lamp that week.

Farms were run quite differently from the croft, of course, though conditions were not necessarily better. My brother Jimmy worked away from home at neighbouring farms, and often found that conditions on the farms were poorer than those at home. At Capenoch, he received for his pay cheques paid to the farmer for the sale of calves, because there was no cash on the farm. Men living in on farms were expected to be single. They were cheaper to hire, and might live above the kitchen, or up the back stair, sharing rooms; cottages were kept for married men. Conditions in the cottages were often primitive. They rarely had bathrooms until the 1950's, and it was only then that farmers installed them in order to attract workers, who were going elsewhere for better housing.

The farmers varied widely in their treatment of the live-in workers; often they were part of the household and the hired men sat down to the same table as the farmer and his family. By the end of the war, however, you could begin to notice an arrogance on the part of the farmers' sons towards the workers. They would wear white coats at the Cattle Show and make efforts to make you feel inferior, though they would often expect to sleep with farm-workers' daughters. Many of them returned to the countryside when war broke out to take up farming jobs which were exempt occupations. Things worsened when the war increased the powers of the farmers over workers, because if you were employed in agriculture and were over the age of 18, you could not move freely. You could be sacked, but you could not leave without notice and a reasonable excuse: my brother Jimmy, who worked at Redbrae and then lived in at Clauchrie, was replaced in favour of a sixteen year old.

Farmers of Ayrshire origin had a reputation for meanness. You could still identify them by their dialect, though they were by this time second generation settlers. It was clear that the older generation of local people disliked them. Often, however, it was simply the case that money had been borrowed by them to set up in a farm, or to buy the dairy cows for it. They were often saving to buy or rent a farm, beginning on smaller farms, and working their way up. They may have worked their men hard, but they worked themselves and their wives equally hard.

On a farm, there was a definite hierarchy: the first ploughman was a man capable of a variety of skilled jobs, not only ploughing, but drilling, building stacks, and had authority over the other workers. He would lead the field in thinning turnips, and he would drive the binder. Beneath him, on a decent size farm, was the second ploughman and the oddman. A favourite shout from the head ploughman on a wet morning to the men assembled in a barn would be: "All you men with coat and leggings, go and stook, and all ye others, go and help them". You were, of course, expected to supply your own waterproofs. Crofters had a lesser status than farmers, but were better off than farm-workers; the cottars often had to take abuse from farmers' sons.

There were a lot more tramps on the road then than now. *Dundee Annie* was around the area at that time. I remember that, at the *Red Lion* in Wigtown, she stole pints of beer from under people's noses and was sometimes thrown out for it. She was generally unpopular, since she was stinking and bad-tempered. *Snib* Scott was the most familiar figure in the Wigtown area. He would come to Torhousemuir begging. Sometimes he would *play* the fiddle, after his own manner. He would get a cup of tea and a *piece*, of which he'd keep half and put half in his pocket. He was fed wherever he went, but occasionally he must have made shift for himself: my brother Jimmy remembers him boiling potatoes in a barrel at Capenoch. Occasionally, he would take rabbit skins, if they were available, to sell. In summer, he would choose a different place every night to sleep, but he had his own favoured places for sleeping in the winter: Barvernochan at Kirkinner was a place where he was welcomed. He had a brother, Tommy, who worked in the smithy at Sorbie. Tommy liked a drink, and I can remember his brandishing a bottle at the cross at Wigtown, during the War, and shouting: "I've stood where thousands fell", and receiving the reply, from a Land Girl nearby: "Aye, when you shook your shirt!" Snib was so well known that he was like part of the countryside. It struck everyone as slightly ridiculous when the Home Guard stopped him and asked for his identity papers. Money didn't seem to motivate him, and he had his own kind of defiant laziness. At Culdarroch, he was asked if he would dig part of the garden, but after he was fed, he replied: "When a man's hungry, he cannae work, and when he's fu', he disnae need tae".

Butcher Murphy was one of the well known local poachers and appears in Ian Niall's books; we knew him because my father used to cut hay from the three fields belonging to the Murphys. He was clever at evading detection, since he would obtain permission to take rabbits, yet Butcher would end up taking them in addition from adjoining farms, where he had no permit. He was once stopped by Sergeant Cannon, when his bicycle was heavily loaded with rabbits. The Sergeant took possession of the bicycle, pushed it up Kirvennie Hill, and down to the *Wash-House* at the Wigtown cross-roads where Butcher lived. It was only at this point that Butcher produced his note, and thanked the Sergeant heartily for his help with the bicycle and went calmly into his house.

The hawkers would come up the Torhousemuir road with their ponies and traps. They would buy rags, sheepskins, scrap iron, and sell from the pack. By the end of the war we probably were buying everything from the Stewarts. The Stewarts sold blanket seconds from the Darvel mills in Ayrshire, where they would go in summer. We favoured them since they were regular callers. My mother also bought drill shirts, bibs and braces, and curtains. When I was married, we purchased a bedding bale from them, and we found their prices were good. There were also tinkers' camps: the Marshalls lived up at Mushy Morton in a tent all year. The mother would come up to Mossend riding a bicycle with a box on the front, and selling needles and thread. Eventually, they moved to a house in the Vennel in Wigtown.

WAR

I WAS nine years old when war began and so I viewed the War from a child's point of view. I remember hearing the Prime Minister announcing that an ultimatum had been given to Germany, failing which we would be at war. I can remember posing many questions about what times of day the fighting stopped and started, and where the food for the troops came from. Our source of information was the wireless, which we had got in about 1937, and when the batteries went down, we almost literally glued our ears to it. We had put the aerial up in the trees above the croft roof for better reception. The accumulator was recharged at Bladnoch and you would get it back, at the cost of sixpence, two days later. We usually listened to it on a Saturday night.

The War had an immediate impact on the manpower in the countryside. You would hear regularly about men receiving *their papers*, which would be the call up to one of the Armed Services. Farm-work was, however, a reserved occupation, and an allocated number of men were exempted on each farm. As I said, farmers' sons often returned home to take up this reserved occupation and there was a good deal of ill-feeling about this. For farm–workers, moving between farms required government permission, and six months' notice. For the first time, subsidies also were introduced: they were for ploughing up land which had not been ploughed for seven years and there was a £10 per acre subsidy for growing potatoes.

Although many men were called up, the countryside suddenly came alive with people. Contractors came to build the camp at Burrow Head, and construct the aerodromes. You would meet wagons on the Newton Stewart road, hauling hardcore for the runways at Baldoon. There was an influx of men and women in uniform: WAAF's, airmen and soldiers. Wigtown was full of off-duty airmen. Trains and buses were full: there were as many standing as there were seated. Some traders cashed in on the situation and might meet troop trains, selling cups of tea. There were Land Girls – some from elsewhere in Scotland, some local and some English. They were billeted at Lochan Croft in Wigtown.

On the estate, soldiers were billeted under canvas outside Torhousemuir House, and the officers lived in the house; others were at Mochrum Park. They were eventually sent to North Africa. On their days off, soldiers would help at the farms, or be sent there compulsorily. Between 1943 and 1946, there were Prisoners of War in the area: Italian Prisoners of War, who were billeted at Minnigaff, worked on the farms in squads of seven or eight people. I can remember their coming turnip-thinning with a guard and a rifle at the end of the rows. Eventually, the regime grew more relaxed and they were allowed to come on their own, or to live in on the farms. Some farmers were meaner than others, and made them take meals in a barn, but others would let them eat with the workers at the farm-house. The majority of them worked hard, and some stayed for good. Collaborators wore khaki uniforms without patches, whereas the Italian Fascists wore a brown uniform with patches. Farms which had requested helpers would get German POW's dropped off at the road-ends; you were not supposed to fraternise with them. There were also *displaced persons:* Ukrainians, Latvians, and Lithuanians. While the POW's were repatriated – eventually - after the war, the DP's could not go home and many stayed and settled.

I only knew of one instance of hostility to a collaborator, and mainly the prisoners steered clear of any trouble. Once, when we were gathering sheep on the moor, I caught a rabbit hiding in a clap and killed it with a stick; I got a second, crossed their paws and hung them through my belt. When I got back to the farm at night, I gave them as a gift to the Italian prisoners. In a day or two, they gave me a cake they had made in the camp, and asked for a template of my mother's finger, in order to make a silver ring for her out of a two-shilling piece. After that, we always gave them rabbits.

Early on in the war, the Local Defence Volunteers established look-out posts on any high hill; as the Home Guard was formed, they took over the manning of the look-outs. Three people would man the post on any given night, for three nights in the month. The tour of duty was for two hours on and two hours off. Lord John Fitzroy from Torhousemuir big house was the Captain and made a habit of trying to catch the sentries unawares. It was while my father was on duty that he saw the flames rising through the trees and engulfing Torhousemuir House. By the time he arrived on the scene, it was too late to save anything. The Home Guard also manned the road-blocks. They set up forty gallon drums about 20 yards apart, forming a chicane. There was one at the junction to the Creamery at Bladnoch Bridge. Lord John Fitzroy refused to stop there when challenged, after he had been drinking at Baldoon Camp, but was rapidly brought to a halt when a bullet went through the back of his windscreen. All the level crossings – at Broughton-Skeog, Gallows Outon and Carslae – were manned.

What affected daily life most dramatically was the black-out. It meant that you had to cover all windows early on in the evening. Even at Mossend, this was strictly observed and enforced. Only the lower half of bicycle lights were left uncovered, and it was the same with cars. When Jimmy worked at Capenoch, he was stopped for having no cover on his rear bicycle lamp, but later stole one from the policeman, who had left his own bike unattended outside the farm. In the country, you could see no light at all, and the complete blackness was a bit daunting: it was as though the countryside were empty. The skylights in the byre were painted blue, and you had to show great caution when opening the front door: the inner door had to be shut so as to prevent the light shining out. A land-mine had been dropped at Barmeal as a result of such carelessness. That, and a German plane which crashed on Cairnsmore was the nearest we came to hostile air attack. Wigtown itself at night was in complete darkness; you might arrive at the pictures in daylight, but by the time you left, the whole town was blacked out. I remember returning from Wigtown after a silver band meeting, and seeing overhead one of the dog-fights which occurred during the Clydeside air-raids. I think we made the journey home to Torhousemuir in record time that night.

At Carsegowan, there was a munitions factory, with its own railway siding. On one occasion, in 1943 there was a huge explosion. I was loading churns at Balmeg, and first saw a huge cloud of debris, and heard the bang seconds later. Two men from Creetown were killed. There were also occasional 'plane crashes at Baldoon: an Anson crashed between Fordbank, near Wigtown, and the cross-roads where you branched off to

Torhousemuir. I saw the crash while cycling from Kirkland. The Canadians who were in the plane are buried at Kirkinner cemetery. They had been practising bombing in a boggy area near Elrig. An Avro Anson also crashed at Carsegowan when it failed to come out of a dive and the pilot's parachute failed to open. He was found in the moss at Carsegowan, and later some time, a glove with the hand still in it turned up.

By the time VE night was celebrated in Wigtown with dancing, drinking, and bonfires everywhere, I was a working man and we had attained my father's ambition and moved from Mossend into a larger croft.

TORHOUSEKIE

IN November 1945, we moved to Torhousekie Croft. We had accumulated a number of young stock in preparation for the move, and our lease was up at Torhousemuir. We had made application to the factor, Mr. Breckenridge, for a tenancy of Torhousekie Croft. It was part of the Mochrum Park estate, owned by Sir James Dunbar. The croft was fifty acres, and the rent, by 1950, was £45 a year.

Torhousekie offered us much larger and what seemed then more luxurious accommodation than Mossend. It might have had no hot water, but it had a pumped supply inside the house, in the back kitchen, which drew on a nearby well. There was an adjacent wash-house and a dry closet outside. The farm steading and dairy was supplied with water at first from a well sunk on Condees Hill, but when this went dry in April of our first year, we fortunately discovered another well when ploughing, and this was connected up to the steading. Mains water did not come until the 1960's, and a bathroom and hot water system were eventually fitted in the 1970's. In the 1950's, farmers began adding bathrooms as lean-to's to farm-cottages to encourage workers to stay in the area, since many had been leaving to seek better housing conditions in the Borders and elsewhere. Kitchens were often added at the same time. Electricity eventually came to the farm cottages in the 1950's.

Torhousekie had a kitchen, with a back-kitchen or pantry. Off the back-kitchen, there was a milk-house with a sandstone floor and sandstone shelves. The sitting room was lit by gas-lighting, from a bottle. Two chains hung down to act as an on or off switch, and you lit a mantle. Upstairs, there were oil-lamps, which were used until the advent of electricity. Off the sitting room, there was a room used as a bedroom, and upstairs were four bedrooms. Willie and I still shared a bedroom, since Jimmy still needed a bedroom at weekends and Jock was home from the War by then. There were five in the house full-time, and, after I left, four, until my father's death in 1968.

Our dairy herd increased to sixteen, only four of which were bought in. We received a gift of two heifer calves from Paton of Auchleand and Willie Lindsay, Carsegowan, whom my father had helped out with work on their farms. There was a new dairy and washing area, with a steam-boiler installed in 1950 by the estate. After this the rent increased to £50 a year. My mother still supervised the milking, though by this time we had a milking machine installed. By the 1950's, Willie saw to the dairying. Eventually, the dairy herd was done away with and the croft went over to beef cattle, because the bulk tanks which came in were not viable for a small herd. When the bulk tanks were emptied by tanker, all the manhandling of the churns which had dictated the routine of Torhousemuir's mornings was gone. You could trace the gradual reduction of the manpower required to run a dairy: first, the milking machine delivered the milk to the churn, and it was carried to the dairy, where it was put into the cooler. This process was replaced by a pipeline. The first milking machines, which consisted simply of a vacuum pump and a petrol-driven engine, only entailed a small outlay. Dairy parlours dispensed with all that and were only possible on large farms.

As I was preparing to leave the countryside, it was undergoing a great period of change. Although tractors had come in as early as 1929, at that time they were few and far between. In the 1940's, however, most machines were being converted to being tractor-drawn. Our horse-drawn binder was converted to fit a tractor draw-bar, and carts acquired pneumatic tyres. The transition from horse-power entailed great changes to the machinery: ploughs, cultivators and other machines had to be renewed or adapted for use by tractors. Most big farms had tractors by 1947 or 1948; by the time I left Wigtownshire in 1950, the horses had largely gone. By that time, we had a tractor with hydraulics: first, a Ferguson, and then, in succession, two Internationals. Blacksmiths changed from shoeing horses to carrying out break-down repairs, and there was also a change from manufacturing parts to merely fitting them. Some blacksmiths became agents for certain makes of machinery, until the bigger companies sent out their own sales representatives. There was a gradual change to machinery powered from the tractor, not from a land-wheel. As the horses disappeared from the farms, so cars gradually appeared on the road. Up to the 1950's, bicycles were still common, but by then farm-workers were saving for cars.

The steam-engines had disappeared from the 1940's, and threshing mills had pneumatic tyres and were pulled by tractors. Combines came in, first pulled by a tractor, then came smaller self-powered versions, and finally much larger machines. At Torhousekie croft, the harvest was still a combined operation of reaper, binder, hay-rake, but these were now tractor-driven. Eventually, a combine was hired in. The threshing was, of course, by this time in the field, not in the stackyard. Ploughing, drilling and cultivating were all carried out by tractor. Muck-spreading became slurry-spreading, so the buildings acquired slatted floors, to enable the slurry to be pumped from beneath them. The Ferguson ploughed two furrows at once; today, tractors plough up to eight, which is the equivalent of the work carried out by sixteen horses. Tractors with cabs came in. Turnips, which had been a staple diet for cattle, were replaced by silage during the winter. In this way, the entire rotation of crops and the cycle of the seasons which we had followed at Torhousemuir was made redundant.

After the War, holdings were split in order to accommodate ex-servicemen and deal with the housing shortage. The Department of Agriculture was landlord of these holdings. There were some at Balfern, and others at Glenluce and Stranraer. Before very long, the process was reversed and these amalgamated again to form larger and more economic units.

My brother John was de-mobbed from the army in 1945, and worked at first on Barness threshing mills. He came home to work in 1950, and he and Willie worked the farm after my father retired in 1948. Willie looked after the farm and the house-work, and Jock did the farm-work. Between 1945 and 1950, I worked on the croft, and worked at the neighbouring farm of Torhousekie. My father died aged 84 in 1968 after a long retirement lived on the croft; my mother died in October 1969. At once point, it looked as if we might have purchased the croft with its land: on the death of Sir James Dunbar, a nephew, Sir Adrian, who had been a house-painter in Maryland, USA, came over and claimed the title. When he began selling off land to realise cash, a meeting was arranged with us to arrange the sale of the croft to my father. When he arrived at the croft, and before the discussions began, he noticed a plaque on the wall displaying the Whiteford coat of arms and a historical note that the Whitefords fought the Dunbars in the fourteenth century. He left in a storm of rage, declaring that he would never speak to us again, and immediately arranged the sale of the land to a neighbouring farmer. My two brothers eventually bought the house in 1970, renovated it, and it is now called *Croft House*.

At Torhousemuir, our croft at Mossend was taken by a man who had been injured in the war, and had part of his face blown away. He stayed five years. It was then taken over by Tom Ronald, who stayed for ten years, and finally moved to the road-end; he worked for the estate. He was the last tenant in Mossend, which was then taken directly under the estate's control and there were no further occupants of the croft. I myself left the area to work in the nuclear industry and did not revisit our croft, by then roofless, until 1993.

As a crofting community, Torhousemuir declined rapidly in the period after we left; eventually, there was virtually only Knockmore which functioned separately from the estate. The Gemmils left just after we did. At Balmeg, Forlow left and had no successor. Harper went out of Hillview, and the McCaigs came in. By 1942/3, Hillend and Hillside became part of Knockmore; Mount Pleasant and Woodside eventually also fused with Knockmore. When the Lindsays left Meadowbank, it became part of the estate. This left the Edmonds, the Horners at Knockskeog, and Knockmore, and the gamekeeper's cottage. Within ten years of our leaving, all the tenants we had known, except the Gemmils and the Horners, were dead. By 1943, the estate had acquired new owners, Sir Archibald and Lady White, who came to Torhousemuir, bringing with them a Mr. Brown, a groom-handyman, who occupied the gardener's cottage, and a lady's maid. Their son, Sir Thomas White, trained as a land-agent, and eventually administered the estate directly. Without tenants, most of the croft buildings became disused, rapidly became derelict, and the moss was planted with conifers.

TORHOUSEMUIR

TORHOUSEMUIR